Fundamentals of CMOS Analog Integrated Circuits

Cosmin Radu Popa

Fundamentals of CMOS Analog Integrated Circuits

Theory and LT Spice Verifications

Springer

Cosmin Radu Popa
National University of Science and Technology Politehnica Bucharest
Bucharest, Romania

ISBN 978-3-032-30676-0 ISBN 978-3-032-30677-7 (eBook)
https://doi.org/10.1007/978-3-032-30677-7

This Springer imprint is published by the registered company Springer Nature Switzerland AG
The registered company address is: Gewerbestrasse 11, 6330 Cham, Switzerland

This book is dedicated to the memory of my mother.

Preface

The design and analysis of analog integrated circuits represent a fundamental component of modern electronics education, yet they are often perceived by students as one of the most challenging areas within the curriculum. This difficulty arises from the intrinsic complexity of semiconductor devices, the non-linear nature of transistor behaviour and the multitude of second-order effects that influence circuit performance in practical technologies. As a consequence, many existing textbooks either emphasize device-level physical modelling in great detail or rely heavily on numerical simulation tools, often at the expense of analytical transparency and intuitive understanding.

In contrast to these approaches, this book is built upon a deliberately simplified, yet rigorous, methodological framework. Its primary objective is to enable students to develop a deep and operational understanding of the fundamental building blocks of CMOS analog integrated circuits through analytical reasoning, supported and validated by circuit simulation. The emphasis is placed not on exhaustive modelling accuracy, but on clarity, consistency and ability to derive meaningful results through systematic hand calculations.

At the core of this approach lies the use of a simplified MOS transistor model. The devices are described using the classical quadratic current-voltage relationship in the saturation region, augmented only by the inclusion of channel-length modulation effect. Other second-order effects (such as body effect, mobility degradation, velocity saturation or short-channel phenomena) are intentionally neglected. While these effects are undoubtedly important in modern deep-submicron technologies, their inclusion would significantly complicate the analytical treatment, often rendering hand calculations impractical or opaque for undergraduate students.

The adoption of this simplified model is not a limitation, but a pedagogical choice. It allows the derivation of closed-form expressions for key circuit parameters, facilitates a clear understanding of the relationships between design variables and supports the development of engineering intuition. At the same time, the inclusion of channel-length modulation effect ensures that essential non-idealities, such as finite output resistance and their impact on voltage gain, are properly captured.

A distinctive feature of this book is the systematic integration of analytical calculations with circuit simulation. All circuits presented throughout the text are implemented and verified using LTspice, with consistent device parameters and supply voltages. The reader is encouraged not only to follow the theoretical derivations, but also to compare them directly with simulation results. This dual perspective serves two important purposes. First, it validates the simplified analytical model by demonstrating its ability to predict circuit behaviour with reasonable accuracy. Second, it highlights the discrepancies that arise as a consequence of modelling assumptions, thereby providing insight into the limitations of first-order analysis.

The content of the book is organized around the most important functional blocks encountered in CMOS analog integrated circuits. Rather than presenting isolated circuits, the material is structured to reflect the hierarchical nature of analog design, where complex systems are built from a relatively small set of fundamental building blocks.

The first chapter introduces current sources and current mirrors, which form the backbone of biasing and signal processing in analog circuits. Both simple and advanced structures are analysed, including cascode and low-voltage wide-swing configurations, as well as self-biased circuits with reduced sensitivity to supply voltage variations. The emphasis is placed on understanding current transfer accuracy, output resistance and impact of device parameters.

The second chapter is devoted to differential amplifiers, which play a central role in nearly all analog integrated circuits. Starting from the basic NMOS and PMOS differential amplifiers with passive load, the analysis develops key concepts such as differential-mode gain, common-mode gain and common-mode rejection ratio. The use of the half-circuit method allows a clear and systematic derivation of these quantities. The study is then extended to differential amplifiers with active loads, which provide significantly improved performance and are widely used in practical designs. Special attention is given to the determination of the allowable maximal range of the common-mode input voltage, a critical aspect for ensuring proper operation.

The third chapter focuses on the internal architectures of CMOS operational amplifiers. Several representative structures are analysed in detail, including basic two-stage or three-stage operational amplifiers, cascode configurations and folded cascode topologies. For each architecture, the operating principle is explained, the DC operating point is established and key performance parameters (such as voltage gain, input and output resistances and maximal range of the common-mode input voltage) are derived analytically. This chapter provides the necessary foundation for understanding how complex analog systems are constructed from simpler building blocks.

The fourth chapter shifts the perspective from transistor-level design to system-level applications. Operational amplifiers are considered as functional blocks, and their behaviour is analysed in feedback configurations. A wide range of applications are covered, including voltage followers, inverting and noninverting amplifiers, summing and difference amplifiers, instrumentation amplifiers and precision rectifiers. Under the assumption of high open-loop gain, the resulting circuit behaviour is shown to depend primarily on ratios of passive components, leading to simple and powerful design relationships. The use of sub-circuit representations enables a seamless connection between the internal architectures studied previously and their external applications.

The book is intended primarily for third-year undergraduate students in electrical engineering and related fields, who have prior exposure to electronic devices and basic circuit theory. It is designed to be accessible to a broad audience, including students specializing in electronics, microelectronics and telecommunications. Given the diversity of backgrounds and the large number of students typically enrolled in such courses, particular care has been taken to ensure that the material remains clear, structured and pedagogically effective.

Compared to existing textbooks, this work distinguishes itself in several important ways. First, it maintains a strict consistency between the analytical model and the simulation environment, allowing direct and meaningful comparisons. Second, it prioritizes analytical transparency, ensuring that all key results can be derived through manageable calculations. Third, it focuses on a carefully selected set of circuits that are sufficiently simple to be fully understood, yet sufficiently representative to illustrate essential design principles.

By combining simplified theory, rigorous analysis and systematic simulation validation, this book aims to provide a solid and coherent foundation for the study of CMOS analog integrated circuits. It equips the reader not only with the ability to analyse and understand fundamental circuits, but also with the conceptual tools necessary for more advanced studies and practical design work in the field of analog electronics.

A key feature of this book is the integration of theoretical analysis with circuit simulation. To support this approach, all circuit schematics used throughout the book are made available to students as pre-designed LTspice files. This allows readers to directly run simulations, explore circuit behaviour and verify the analytical results presented in the text. By providing ready-to-use simulation files, the book enables a hands-on learning experience without requiring students to spend time on schematic entry, allowing them instead to focus on understanding circuit operation and validating theoretical predictions. As a result, the book is intended to serve as a comprehensive support for lectures, seminars and laboratory sessions in the course on analog integrated circuits, offering a unified framework that combines theory, analysis and practical verification.

Bucharest, Romania Cosmin Radu Popa

Acknowledgements

This book was funded by the project "National Platform for Semiconductor Technologies", contract no. G 2024-85828/390008/27.11.2024, SMIS code 351364, funded by the European Regional Development Fund under the Program for Intelligent Growth, Digitization, and Financial Instruments.

Competing Interests The author has no competing interests to declare that are relevant to the content of this manuscript.

Contents

cascode configurations and self-biased structures, focusing on their theoretical foundations and practical implementation aspects.

The analysis begins with the basic current sources, where the simplicity of the topology allows a clear understanding of how device parameters such as output resistance or minimal output voltage influence current stability. Then, the discussion extends to the cascode current sources, which improves output resistance, in order to reduce the effect of channel-length modulation. Finally, self-biased current sources are introduced, emphasizing their ability to generate a reference current with an extremely low dependence on the supply voltage variations.

Each structure is analysed using the classical square-law MOS model, allowing a direct connection between theoretical expressions and simulation results. LTspice simulations accompany the analytical development, in order to validate the theoretical conclusions under controlled conditions. This approach offers students a clear and consistent framework for understanding the operation, limitations and design trade-offs of MOS current sources in analog integrated circuits.

All LTspice simulation files associated with the circuits presented in this book are provided as supplementary material on the publisher's website. Each file is named identically to the corresponding figure in the chapter, with the extension ".asc". The reader is encouraged to run, analyse and modify these files in order to validate the theoretical results and explore circuit behavior. The use of these materials requires the installation of the LTspice simulation software.

1.2 The Model of the Current Source

The most convenient method for modelling a real current source uses two components: an ideal current generator I_O and a parallel-connected resistor R_O—the output resistance of the current source (Fig. 1.1a). I_O current is perfectly constant, so it does not depend on the voltage across the current source V_O, on supply voltage, on temperature variations or on technologically caused errors. The output voltage of the current source V_O is an externally applied voltage, imposed by the circuit that uses the current generated by the current source. So, the current source itself is not able to set the output voltage at which it operates; only the external circuit can do this. Because the current source is implemented using transistors that must be properly biased, there exists a minimal output voltage for any current source circuit V_{Omin}.

The graphical representation of a general output characteristic of a current source is shown in Fig. 1.1b.

In order to quantitatively evaluate the performances of a current source, the most important parameters are the following:

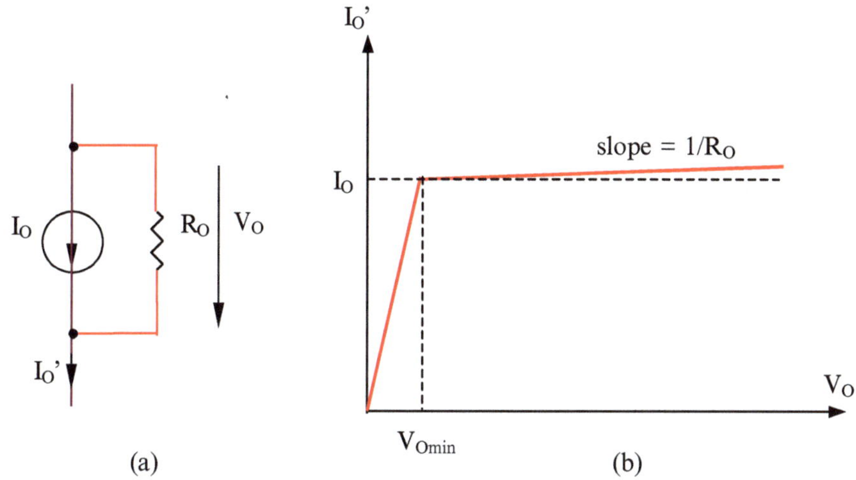

Fig. 1.1 (**a**) The model of a current source; (**b**) the general output characteristic $I_O^{'}(V_O)$ of a current source

1.2.1 Output Current $\mathbf{I_O}$

The output current of a current source can be expressed as a sum of two components: a constant current I_O and another component that depends on the output voltage V_O (applied to the current source by the external circuit):

$$I_O{}' = I_O + \frac{V_O}{R_O}. \tag{1.1}$$

This relation shows that the output current of a current source is linearly increasing when the output voltage applied to it increases. The slope of the mathematical characteristic $I_O^{'}(V_O)$ (named **output characteristic of the current source)** is equal to $1/R_O$.

1.2.2 Output Resistance $\mathbf{R_O}$

The output resistance of a current source is defined as follows:

$$R_O = \left(\frac{dI_O}{dV_O}\right)^{-1}. \tag{1.2}$$

It is equal to the inverse of the slope of the output characteristic from Fig. 1.1b. In order to minimize the dependence of the output current $I_O^{'}$ on the output voltage V_O

variations, the output resistance of the current source must be as large as possible (see Eq. (1.1)). An ideal current source has an infinite value of the output resistance R_O.

1.2.3 *Minimal Output Voltage* V_{Omin}

The minimal output voltage of a current source is the minimal value of the voltage applied between the terminals of a current source, which allows a correct operation of the current source. For maximizing the range in which the current source can be used, the minimal output voltage must be as small as possible.

1.2.4 *Power Supply Rejection Ratio (PSRR)*

This parameter evaluates the dependence of the output current I_O on the supply voltage variations:

$$PSRR = \frac{dI_O}{dV_{DD}}. \tag{1.3}$$

1.2.5 *Sensitivity of the Output Current on the Supply Voltage Variations* $S_{I_O}^{V_{DD}}$

This parameter is practically equivalent to *PSRR*, but it has the advantage of being non-dimensional. For this reason, it simplifies the comparison between two current sources from the point of view of the dependence of their output currents on supply voltage:

$$S_{I_O}^{V_{DD}} = \frac{dI_O/I_O}{dV_{DD}/V_{DD}} = \frac{V_{DD}}{I_O}\frac{dI_O}{dV_{DD}} = \frac{V_{DD}}{I_O}PSRR. \tag{1.4}$$

1.2.6 Temperature Coefficient (TC)

The temperature coefficient evaluates the dependence on temperature of the output current. This behavior is a consequence of the temperature variations of multiple circuit parameters (usually modelled by complex mathematical functions):

$$TC = \frac{dI_O}{dT}. \tag{1.5}$$

1.2.7 Relative Temperature Coefficient (RTC)

This parameter is equivalent to *TC*, being defined as follows:

$$RTC = \frac{1}{T}\frac{dI_O}{dT} = \frac{1}{T}TC. \tag{1.6}$$

1.3 Basic Current Sources

1.3.1 General MOS Current Source (Version I)

The simplest implementation of a current source uses an MOS transistor M and a voltage source V_1 (that must be implemented using a particular circuit, for obtaining the full schematic of the current source). The circuit of the current source is shown in Fig. 1.2.

1.3.1.1 Output Current Expression I_O

The output current I_O for this current source can be expressed using the squaring characteristic of transistor M, biased in saturation region:

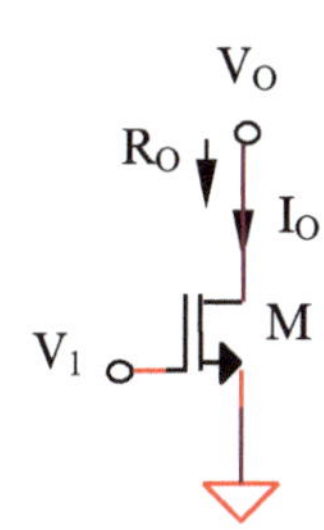

Fig. 1.2 General MOS current source (version I)

$$I_O = I_D = \frac{K'_n}{2}\left(\frac{W}{L}\right)(V_{GS} - V_T)^2(1 + \lambda_n V_{DS}). \tag{1.7}$$

Because the gate-source voltage of transistor M is imposed by the voltage source V_1 and its drain-source voltage is equal to the output voltage V_O, the previous relation becomes

$$I_O = \frac{K'_n}{2}\left(\frac{W}{L}\right)(V_1 - V_T)^2(1 + \lambda_n V_O). \tag{1.8}$$

The output current of the current source from Fig. 1.2 is mainly determined by the biasing voltage V_1 and by the parameters of transistor M. The channel-length modulation effect that affects the operation of transistor M introduces a small linear increasing of the output current I_O when the output voltage V_O increases. In order to fulfil a biasing in saturation of transistor M, the value of voltage V_1 must be greater than the threshold voltage of the transistor V_T (because $V_1 = V_{GS1}$).

1.3.1.2 Output Resistance Expression R_O

Considering the low-frequency small-signal model of the MOS transistor (determined using Fig. A.1 from Annex), the output resistance of the current source from Fig. 1.2 can be expressed as follows:

$$R_O = r_{ds} = \frac{1}{\lambda_n I_D} = \frac{1}{\lambda_n I_O}. \tag{1.9}$$

1.3.1.3 Minimal Output Voltage V_{Omin}

In order to bias transistor M in saturation, the following fundamental condition must be fulfilled:

$$V_{DS} \geq V_{GS} - V_T. \tag{1.10}$$

Because the gate-source voltage of transistor M is equal to V_1 voltage and the output voltage of the MOS current source V_O is equal to V_{DS}, the minimal output voltage will have the following expression:

$$V_{O\,min} = V_{GS2} - V_T. \tag{1.11}$$

By neglecting the channel-length modulation that affects the operation of transistor M, the squaring characteristic of this transistor can be expressed as follows:

$$I_O \cong \frac{K'_n}{2}\left(\frac{W}{L}\right)(V_{GS}-V_T)^2 = \frac{K'_n}{2}\left(\frac{W}{L}\right)(V_1-V_T)^2. \tag{1.12}$$

From the previous two relations, another equivalent mathematical expression of the minimal output voltage results:

$$V_{O\,min} \cong \sqrt{\frac{2I_O}{K'_n\left(\frac{W}{L}\right)}}. \tag{1.13}$$

1.3.1.4 Design Examples

In order to confirm the previously presented theoretical estimated results, the MOS current source from Fig. 1.2 is simulated with LTspice using the following values: $K'_n = 200\,\mu A/V^2$, $\lambda_n = 0.01\,V^{-1}$ and $V_{Tn} = 0.5\,V$. MOS transistor has the aspect ratio $\left(\frac{W}{L}\right) = \frac{5\,\mu m}{0.5\,\mu m} = 10$. The value of the voltage source V_1 is equal to 1 V.

The theoretical values of the most important parameters of the current source from Fig. 1.3 could be determined using Eqs. (1.7), (1.9) and (1.11), respectively:

$$I_O \cong \frac{K'_n}{2}\left(\frac{W}{L}\right)(V_{GS}-V_T)^2 = \frac{200\,\mu A/V^2}{2} \times 10 \times (1\,V - 0.5\,V)^2 = 250\,\mu A, \tag{1.14}$$

$$R_O = \frac{1}{\lambda_n I_O} = \frac{1}{0.01\,V^{-1} \times 250\,\mu A} = 400\,k\Omega \tag{1.15}$$

and

$$V_{O\,min} = V_1 - V_T = 1\,V - 0.5\,V = 0.5\,V. \tag{1.16}$$

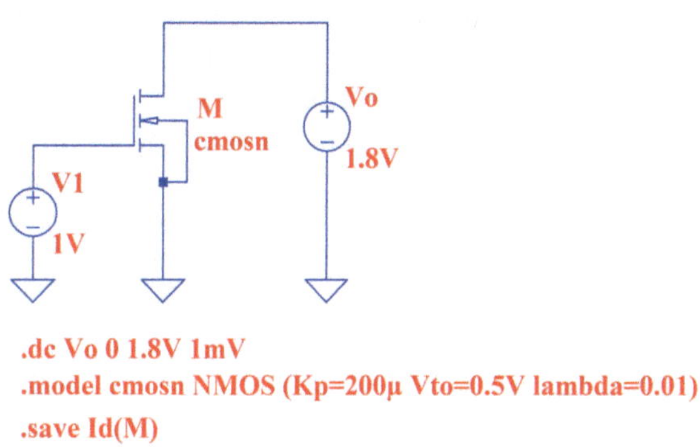

Fig. 1.3 MOS current source with *DC sweep analysis* for output voltage V_O

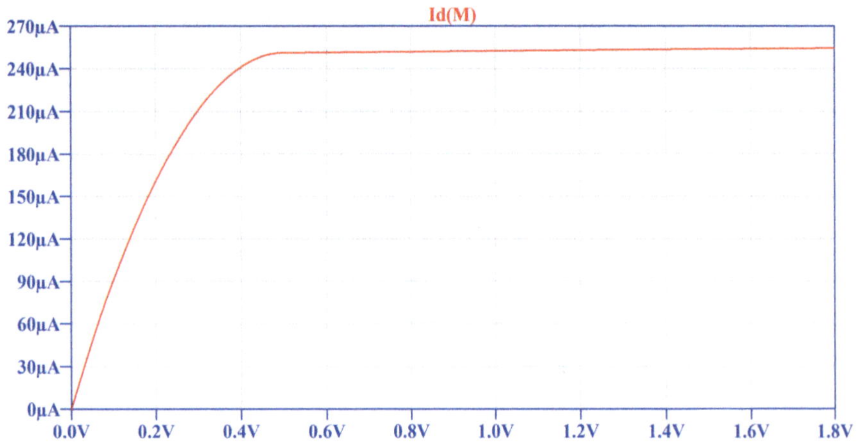

Fig. 1.4 Output characteristic $I_O(V_O)$ for the current source presented in Fig. 1.3

The circuit used for simulating the output characteristic $I_O(V_O)$ of the general MOS current source (version I) is shown in Fig. 1.3.

Setting a variation of the output voltage V_O between 0 and 1.8 V (a *DC sweep analysis* is performed in LTspice), the graphical representation of the output current I_O is shown in Fig. 1.4 (the output characteristic of the MOS current source from Fig. 1.3). In order to directly view the result of the simulation (characteristic $I_O(V_O)$, shown in Fig. 1.4), the directive ".save Id(M)" is introduced in the file, with Id (M) representing the output current of the circuit I_O.

The simulated value of the output current I_O is approximately 251 μA, very close to the theoretical value of 250 μA, expressed by Eq. (1.14).

The output current I_O is linearly increasing with the output voltage, but only for an output voltage superior to or equal to the minimal output voltage V_{Omin}, expressed by Eq. (1.16).

For a more facile evaluation with LTspice of the output resistance of the MOS current source from Fig. 1.3, the range of variation for the output voltage V_O is limited (a *DC sweep analysis* is performed in LTspice, with the variation range for the output voltage V_O being set to be between 0.4 and 1.8 V). The circuit used for this simulation is presented in Fig. 1.5, with the graphical representation of the output current I_O being shown in Fig. 1.6. In order to directly view the result of the simulation, the directive ".save Id(M)" is introduced in the file, with Id (M) representing the output current of the circuit I_O.

In this simulation, the slope of the $I_O(V_O)$ characteristic is equal to $1/R_O$, with R_O being the output resistance of the MOS current source from Fig. 1.5. In consequence, the output resistance will have the following value:

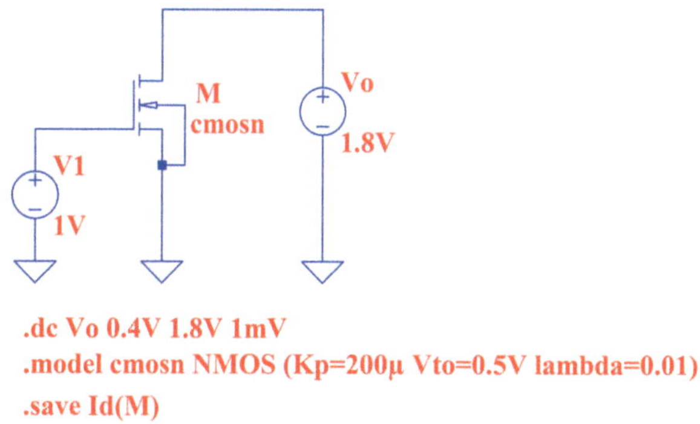

Fig. 1.5 MOS current source with limited range of the *DC sweep analysis* for output voltage variation

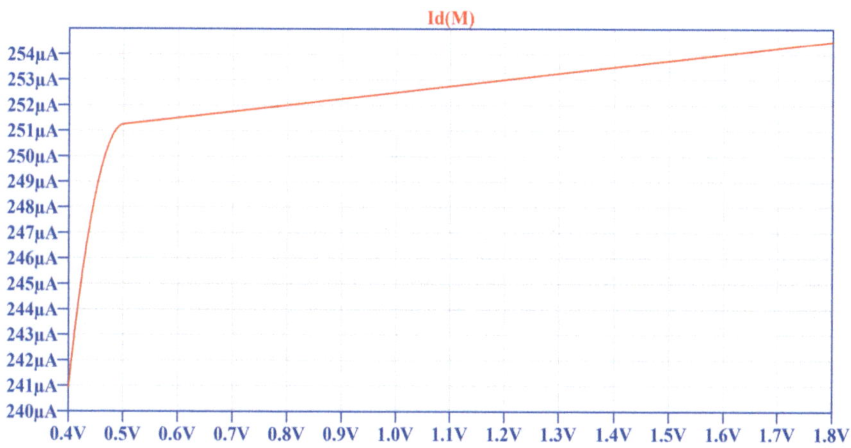

Fig. 1.6 Graphical representation of the output current I_O as a function of the output voltage V_O, for a limited range of the output voltage

$$R_O = \frac{1.8\,\text{V} - 0.8\,\text{V}}{254.5\,\mu\text{A} - 252\,\mu\text{A}} = \frac{1\,\text{V}}{2.5\,\mu\text{A}} = 400\,\text{k}\Omega, \tag{1.17}$$

identical with the value of the theoretical output resistance, computed in Eq. (1.15). This value is obtained for a channel length equal to 0.5 μm. In order to increase the value of the output resistance R_O, the channel length of transistor M, L, can be increased, maintaining the value of its aspect ratio (W/L). This will reduce the value of the channel-length modulation parameter λ_n and, in consequence, will increase the value of the output resistance R_O. The disadvantage of this method is that it strongly increases the silicon-occupied area. For example, the utilization of a double-channel length of transistor M will increase four times the silicon-occupied area of the circuit (from $0.5\,\mu\text{m} \times 5\,\mu\text{m} = 2.5\,\mu\text{m}^2$ to $1\,\mu\text{m} \times 10\,\mu\text{m} = 10\,\mu\text{m}^2$, for the same $(W/L) = 10$).

The simulated value of the minimal output voltage is equal to 0.5 V, identical with the theoretical value of V_{Omin}, expressed by Eq. (1.16).

Fig. 1.7 Circuit used for obtaining the graphical representation of $I_O(V_O)$ characteristics for multiple values of channel-length parameter λ_n

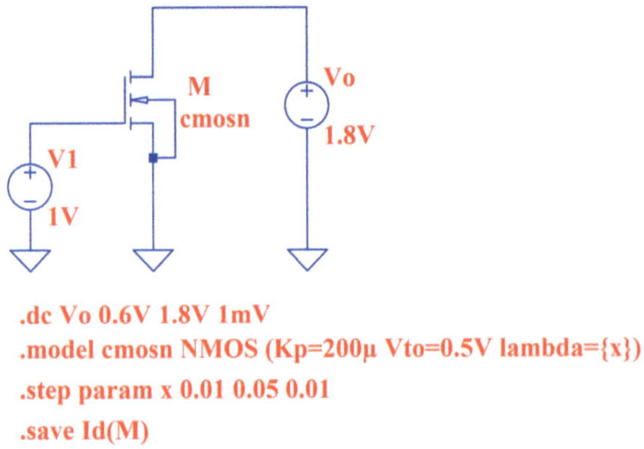

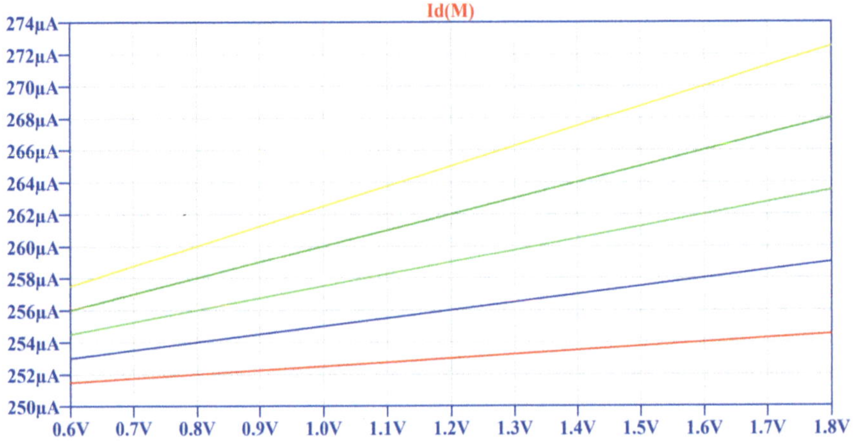

Fig. 1.8 Graphical representation of the characteristics $I_O(V_O)$, for multiple values of the channel-length parameter λ_n

In order to graphically represent the dependence of the output current I_O on the output voltage V_O, for multiple values of channel-length parameter λ_n (0.01, 0.02, 0.03, 0.04 and 0.05 V^{-1}), a *DC sweep analysis* is performed in LTspice for a range of the output voltage V_O between 0.6 and 1.8 V and also a parametrical simulation for $x = \lambda_n$, between 0.01 and 0.05 V^{-1}, with a step of 0.01 V^{-1}. The circuit used for this simulation is presented in Fig. 1.7.

The parametrical simulation results for the output characteristics of the current source $I_O(V_O)$ are illustrated in Fig. 1.8. In order to directly view the results of the simulation, the directive ".save Id(M)" is introduced in the file, with Id (M) representing the output current of the circuit I_O.

In Fig. 1.8, the line with the greatest slope represents the simulation for the largest value of channel-length modulation parameter $\lambda_n = 0.05\ \text{V}^{-1}$, equivalent to the shortest channel length of transistor M and to the smallest value of the output resistance R_O for the MOS current source from Fig. 1.7.

The line with the smallest slope represents the simulation for the smallest value of channel-length modulation parameter $\lambda_n = 0.01\,\mathrm{V}^{-1}$, equivalent to the largest channel length of transistor M and to the largest value of the output resistance R_O for the MOS current source from Fig. 1.7.

1.3.2 General MOS Current Source (Version II)

In order to introduce an additional possibility of controlling the output current of the current source from Fig. 1.2 and in order to increase its output resistance, an additional resistor can be introduced between the source of transistor M and ground, resulting in the current source from Fig. 1.9. The voltage source V_1 must be implemented using a particular circuit, for obtaining the full schematic of the current source.

1.3.2.1 Output Current Expression I_O

The expression of V_1 voltage can be written as follows:

$$V_1 = RI_O + V_{GS}. \tag{1.18}$$

Because of the biasing in saturation of transistor M, the output current can be approximately expressed as

$$I_O \cong \frac{K'_n}{2}\left(\frac{W}{L}\right)(V_{GS} - V_T)^2. \tag{1.19}$$

In order to determine the expression of current I_O, the system of equations containing previous two relations must be solved. Because we have the $V_{GS} \geq V_T$

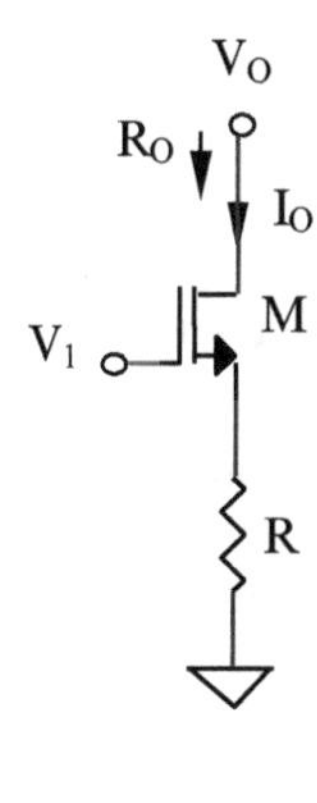

Fig. 1.9 General MOS current source (version II)

criteria available for choosing the correct solution of the second-order equation that will be obtained by analysing these relations (because of the biasing in saturation region of transistor M), the expression of current I_O from Eq. (1.19) must be replaced in Eq. (1.18), resulting in the following second-order equation:

$$V_1 = R\frac{K'_n}{2}\left(\frac{W}{L}\right)(V_{GS} - V_T)^2 + V_{GS}. \tag{1.20}$$

The solutions of this equation are

$$(V_{GS})_A = V_T + \frac{1}{K'_n\left(\frac{W}{L}\right)R}\left(\sqrt{1 + 2K'_n\left(\frac{W}{L}\right)R(V_1 - V_T)} - 1\right) \tag{1.21}$$

and

$$(V_{GS})_B = V_T - \frac{1}{K'_n\left(\frac{W}{L}\right)R}\left(\sqrt{1 + 2K'_n\left(\frac{W}{L}\right)R(V_1 - V_T)} + 1\right). \tag{1.22}$$

Because only $(V_{GS})_A > V_T$, it will represent the mathematical solution of Eq. (1.20), resulting in the following expression of the output current:

$$I_O = \frac{1}{K'_n\left(\frac{W}{L}\right)R^2}\left[K'_n\left(\frac{W}{L}\right)R(V_1 - V_T) - \sqrt{1 + 2K'_n\left(\frac{W}{L}\right)R(V_1 - V_T)} + 1\right]. \tag{1.23}$$

1.3.2.2 Output Resistance Expression R_O

Considering the expression of the dynamic resistance viewed in the drain terminal of the MOS transistor (determined using Eq. (A.10) from Annex), the output resistance of the current source from Fig. 1.9 can be expressed as follows:

$$R_O = r_{ds}(1 + g_m R) = \frac{1}{\lambda_n I_D}(1 + g_m R) = \frac{1}{\lambda_n I_O}(1 + g_m R). \tag{1.24}$$

1.3.2.3 Minimal Output Voltage V_{Omin}

In order to bias transistor M in saturation, the fundamental condition $V_{DS} \geq V_{GS} - V_T$ must be fulfilled. It results in the following expression of the minimal output voltage of the current source from Fig. 1.9:

$$V_{O\,min} = V_{GS} - V_T + I_O R. \tag{1.25}$$

By neglecting the channel-length modulation effect that affects the operation of transistor M, the squaring characteristic of this transistor can be expressed as follows:

$$I_O \cong \frac{K'_n}{2}\left(\frac{W}{L}\right)(V_{GS} - V_T)^2. \tag{1.26}$$

From the previous two relations, another equivalent mathematical expression of the minimal output voltage results:

$$V_{O\,min} \cong \sqrt{\frac{2I_O}{K'_n\left(\frac{W}{L}\right)}} + I_O R. \tag{1.27}$$

1.3.2.4 Design Examples

In order to confirm the previously presented theoretical estimated results, the MOS current source from Fig. 1.9 is simulated with LTspice using the following values: $K'_n = 200\,\mu\text{A/V}^2$, $\lambda_n = 0.01\,\text{V}^{-1}$ and $V_{Tn} = 0.5\,\text{V}$. MOS transistor has the aspect ratio $\left(\frac{W}{L}\right) = \frac{5\,\mu\text{m}}{0.5\,\mu\text{m}} = 10$. The value of the voltage source V_1 is equal to 1 V, and the value of the resistor is $R = 1\,\text{k}\Omega$.

The theoretical values of the output current for the current source from Fig. 1.9 could be determined using Eqs. (1.19) and (1.20):

$$\frac{200\,\mu\text{A/V}^2}{2} \times 10 \times 1\,\text{k}\Omega \times (V_{GS} - 0.5\,\text{V})^2 + V_{GS} = 1\,\text{V}, \tag{1.28}$$

resulting in the value $V_{GS} = 0.866\,\text{V}$, corresponding to a biasing in saturation region of transistor M. Replacing this value in Eq. (1.19) results in

$$I_O \cong \frac{200\,\mu\text{A/V}^2}{2} \times 10 \times (0.866\,\text{V} - 0.5\,\text{V})^2 = 133.96\,\mu\text{A}. \tag{1.29}$$

The theoretical value of the output resistance can be obtained using Eq. (1.24). The transconductance g_m has the following value:

$$g_m = \sqrt{2K'_n\left(\frac{W}{L}\right)I_O} = \sqrt{2 \times 200\,\mu\text{A/V}^2 \times 10 \times 133.96\,\mu\text{A}} = 732\,\mu\text{A/V}, \tag{1.30}$$

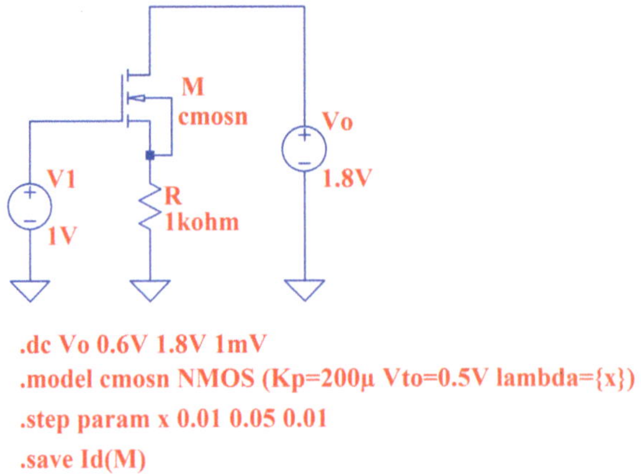

Fig. 1.14 Circuit used for obtaining the graphical representation of the characteristics $I_O(V_O)$, for multiple values of channel-length parameter λ_n

simulation, the directive ".save Id(M)" is introduced in the file, with Id (M) representing the output current of the circuit I_O.

In this simulation, the slope of the $I_O(V_O)$ characteristic is equal to $1/R_O$, with R_O being the output resistance of the MOS current source from Fig. 1.12. In consequence, the output resistance will have the following value:

$$R_O = \frac{1.8\,\text{V} - 0.8\,\text{V}}{135.26\,\mu\text{A} - 134.49\,\mu\text{A}} = \frac{1\,\text{V}}{0.77\,\mu\text{A}} = 1.298\,\text{M}\Omega, \tag{1.33}$$

very close to the value of the theoretical output resistance, computed in Eq. (1.31).

The simulated value of the minimal output voltage is equal to 0.5 V, identical with the theoretical value of V_{Omin}, expressed by Eq. (1.32).

In order to graphically represent the dependence of the output current I_O on the output voltage variations, for multiple values of channel-length parameter λ_n (0.01, 0.02, 0.03, 0.04 and 0.05 V^{-1}), a *DC sweep analysis* is performed in LTspice for a range of the output voltage V_O between 0.6 and 1.8 V and also a parametrical simulation for $x = \lambda_n$, between 0.01 and 0.05 V^{-1}, with a step of 0.01 V^{-1}. The circuit used for this simulation is presented in Fig. 1.14.

The parametrical simulation results for the output characteristics of the current source $I_O(V_O)$ are illustrated in Fig. 1.15. In order to directly view the result of the simulation, the directive ".save Id(M)" is introduced in the file, with Id (M) representing the output current of the circuit I_O.

In Fig. 1.15, the line with the greatest slope represents the simulation for the largest value of the channel-length modulation parameter $\lambda_n = 0.05\,\text{V}^{-1}$, equivalent to the shortest channel length of transistor M and to the smallest value of the output resistance R_O for the MOS current source from Fig. 1.14.

The line with the smallest slope represents the simulation for the smallest value of the channel-length modulation parameter $\lambda_n = 0.01\,\text{V}^{-1}$, equivalent to the largest channel length of transistor M and to the largest value of the output resistance R_O for the MOS current source from Fig. 1.14.

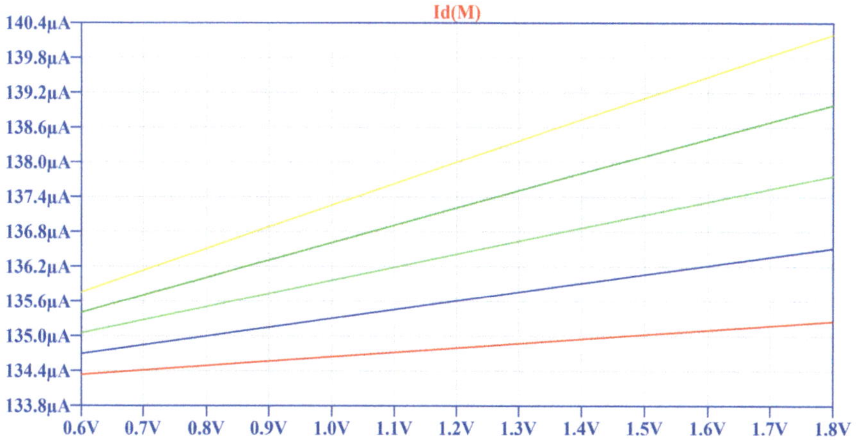

Fig. 1.15 Graphical representation of the characteristics $I_O(V_O)$, for multiple values of the channel-length parameter λ_n

Fig. 1.16 Basic MOS current mirror

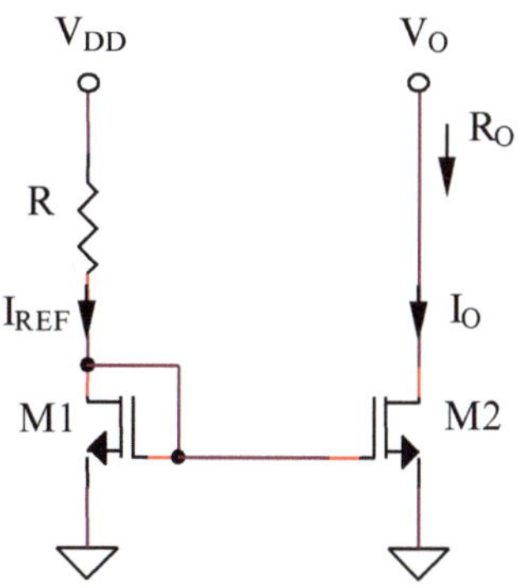

1.3.3 Basic MOS Current Mirror

The basic MOS current mirror is one of the fundamental building blocks in analog circuit design, being extensively treated in the classical textbooks of analog integrated circuits [1–4]. It is used to generate precise, mirrored current, which can be replicated and delivered to multiple branches of a circuit, ensuring currents that are stable and matched over a wide range of conditions. The most important function of a basic MOS current mirror is to take an input reference current I_{REF} and to mirror it across multiple output branches, effectively duplicating the current with reasonable accuracy.

In practical applications, current mirrors are used in biasing circuits, amplifiers, voltage references and current sources, serving as an important component for stable operation of many analog systems.

The circuit of the basic MOS current mirror is illustrated in Fig. 1.16.

1.3.3.1 Output Current Expression I_O

The expression of the supply voltage can be written as follows:

$$V_{DD} = RI_{REF} + V_{GS1}. \tag{1.34}$$

Because of the biasing in saturation of transistor M1, the reference current can be expressed as

$$I_{REF} \cong \frac{K'_n}{2}\left(\frac{W}{L}\right)_1 (V_{GS1} - V_T)^2. \tag{1.35}$$

In order to determine the expression of I_{REF} current, the system of equations containing previous two relations must be solved. Because we have the $V_{GS1} \geq V_T$ criteria available for choosing the correct solution of the second-order equation that will be obtained by analysing these relations (because of the biasing in saturation region of transistor M1), the expression of I_{REF} current from Eq. (1.35) must be replaced in Eq. (1.34), resulting in the following second-order equation:

$$V_{DD} = R\frac{K'_n}{2}\left(\frac{W}{L}\right)_1 (V_{GS1} - V_T)^2 + V_{GS1}. \tag{1.36}$$

The solutions of this equation are

$$(V_{GS1})_A = V_T + \frac{1}{K'_n\left(\frac{W}{L}\right)_1 R}\left(\sqrt{1 + 2K'_n\left(\frac{W}{L}\right)_1 R(V_{DD} - V_T)} - 1\right) \tag{1.37}$$

and

$$(V_{GS1})_B = V_T - \frac{1}{K'_n\left(\frac{W}{L}\right)_1 R}\left(\sqrt{1 + 2K'_n\left(\frac{W}{L}\right)_1 R(V_{DD} - V_T)} + 1\right). \tag{1.38}$$

Because only $(V_{GS1})_A \geq V_T$, it will represent the mathematical solution of Eq. (1.36), resulting in the following expression of the reference current:

$$I_{REF} = \frac{1}{K'_n\left(\frac{W}{L}\right)_1 R^2} \times \left[K'_n\left(\frac{W}{L}\right)_1 R(V_{DD} - V_T) - \sqrt{1 + 2K'_n\left(\frac{W}{L}\right)_1 R(V_{DD} - V_T)} + 1\right]. \tag{1.39}$$

In order to determine the expression of the output current I_O and also the mirroring errors of the basic MOS current mirror from Fig. 1.16, it is useful to compute the ratio I_O/I_{REF}:

$$\frac{I_O}{I_{REF}} = \frac{\frac{K'_n}{2}\left(\frac{W}{L}\right)_2 (V_{GS2} - V_T)^2 (1 + \lambda_n V_{DS2})}{\frac{K'_n}{2}\left(\frac{W}{L}\right)_1 (V_{GS1} - V_T)^2 (1 + \lambda_n V_{DS1})}. \quad (1.40)$$

The quantitative evaluation of the mirroring errors imposes the necessity of taking into account the channel-length modulation effect that affects the MOS transistor operation. Because the circuit of the basic MOS current mirror imposes $V_{GS1} = V_{GS2}$, $V_{DS1} = V_{GS1}$ and $V_{DS2} = V_O$, the previous ratio can be expressed as follows:

$$\frac{I_O}{I_{REF}} = \frac{\left(\frac{W}{L}\right)_2}{\left(\frac{W}{L}\right)_1} \frac{1 + \lambda_n V_O}{1 + \lambda_n V_{GS1}}, \quad (1.41)$$

resulting in

$$I_O = I_{REF} \frac{\left(\frac{W}{L}\right)_2}{\left(\frac{W}{L}\right)_1} \frac{1 + \lambda_n V_O}{1 + \lambda_n V_{GS1}}, \quad (1.42)$$

with I_{REF} expressed by Eq. (1.39).

In consequence, the value of the output current of the fundamental MOS current mirror I_O will slightly increase when the output voltage V_O increases. The magnitude of this undesired behavior of the current mirror mainly depends on the technological parameter λ_n and also on the value of the output voltage V_O at the imposed DC operating point. Neglecting the errors introduced by the channel-length modulation effect, the expression of the output current I_O can be approximately written as

$$I_O \cong I_{REF} \frac{\left(\frac{W}{L}\right)_2}{\left(\frac{W}{L}\right)_1}. \quad (1.43)$$

The simplest implementation of the basic MOS current mirror uses identical MOS transistors, resulting in $I_O \cong I_{REF}$.

1.3.3.2 Output Resistance Expression R_O

Considering the small-signal model of the MOS transistor (determined using Fig. A.1 from Annex), the output resistance of the basic MOS current mirror can be expressed as follows:

$$R_O = r_{ds2} = \frac{1}{\lambda_n I_{D2}} = \frac{1}{\lambda_n I_O}. \tag{1.44}$$

1.3.3.3 Minimal Output Voltage V_{Omin}

In order to bias in saturation both MOS transistors of the current mirror, the fundamental condition $V_{DS} \geq V_{GS} - V_T$ must be fulfilled. For transistor M1, this condition becomes

$$V_{DS1} = V_{GS1} > V_{GS1} - V_T. \tag{1.45}$$

So, transistor M1 is always biased in saturation region. The same condition for transistor M2 could be written as follows:

$$V_{DS2} \geq V_{GS2} - V_T. \tag{1.46}$$

Because the output voltage of the basic MOS current mirror V_O is equal to V_{DS2}, the minimal output voltage of the circuit will have the following expression:

$$V_{O\,min} = V_{GS2} - V_T. \tag{1.47}$$

By neglecting the channel-length modulation effect that affects the operation of transistor M2, the squaring characteristic of this transistor can be expressed as

$$I_O \cong \frac{K_n'}{2}\left(\frac{W}{L}\right)_2 (V_{GS2} - V_T)^2. \tag{1.48}$$

Expressing the overdrive voltage $V_{GS2} - V_T$ from the previous relation and introducing it in the expression of the minimal output voltage, another equivalent mathematical expression of this parameter of the current source results:

$$V_{O\,min} \cong \sqrt{\frac{2I_O}{K_n'\left(\frac{W}{L}\right)_2}}. \tag{1.49}$$

1.3.3.4 Design Examples

In order to confirm the previously presented theoretical estimated results, the basic MOS current mirror from Fig. 1.16 is simulated with LTspice using the following values: $K_n' = 200\,\mu\text{A/V}^2$, $\lambda_n = 0.01\,\text{V}^{-1}$, $V_{Tn} = 0.5\,\text{V}$ and $V_{DD} = 1.8\,\text{V}$. MOS

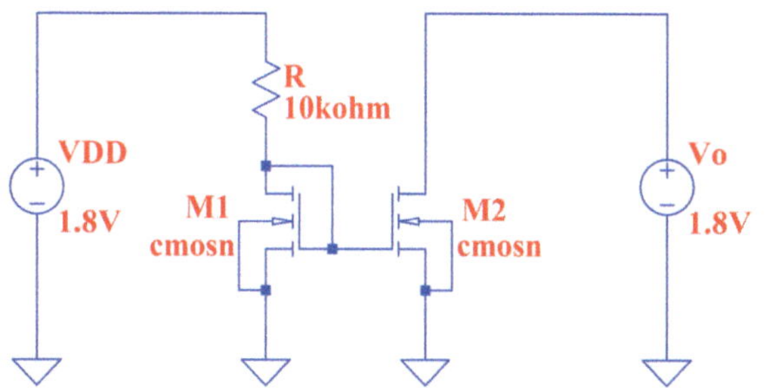

Fig. 1.17 Basic MOS current mirror with a *DC sweep analysis* for the output voltage V_O

transistors have the aspect ratios $\left(\frac{W}{L}\right)_1 = \left(\frac{W}{L}\right)_2 = \frac{5\,\mu\text{m}}{0.5\,\mu\text{m}} = 10$, while the value of the resistor is $R = 10\,\text{k}\Omega$.

The theoretical value of the output current for the basic MOS current mirror from Fig. 1.16 could be determined using Eq. (1.36):

$$\frac{200\,\mu\text{A}/\text{V}^2}{2} \times 10 \times 10\text{k}\Omega \times (V_{GS1} - 0.5\,\text{V})^2 + V_{GS1} = 1.8\,\text{V}, \tag{1.50}$$

resulting in the value $V_{GS1} = 0.814\,\text{V}$, corresponding to a biasing in saturation region of transistor M1. Replacing this value in Eq. (1.35) results in

$$I_O \cong \frac{200\,\mu\text{A}/\text{V}^2}{2} \times 10 \times (0.814\,\text{V} - 0.5\,\text{V})^2 = 98.6\,\mu\text{A}. \tag{1.51}$$

The theoretical value of the output resistance can be obtained using Eq. (1.44):

$$R_O = \frac{1}{0.01\,\text{V}^{-1} \times 98.6\,\mu\text{A}} = 1.014\,\text{M}\Omega. \tag{1.52}$$

The theoretical value of the minimal output voltage can be obtained using Eq. (1.47), replacing V_{GS2} with V_{GS1}:

$$V_{O\,min} = 0.814\,\text{V} - 0.5\,\text{V} = 0.314\,\text{V}. \tag{1.53}$$

The circuit used for simulating the output characteristic $I_O(V_O)$ of the basic current mirror is shown in Fig. 1.17.

Setting a variation of the output voltage V_O between 0 and 1.8 V (a *DC sweep analysis* is performed in LTspice), the graphical representation of the output current I_O is shown in Fig. 1.18 (the output characteristic of the basic MOS current mirror from Fig. 1.17). In order to directly view the result of the simulation, the directive ".

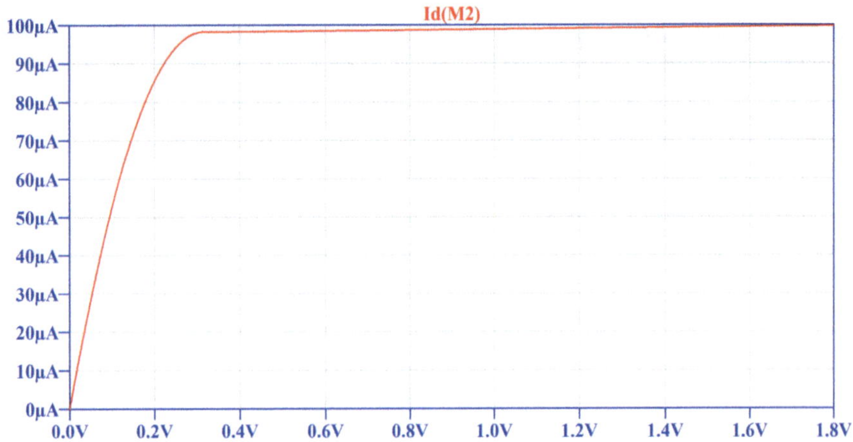

Fig. 1.18 Output characteristic $I_O(V_O)$ for the basic MOS current mirror presented in Fig. 1.17

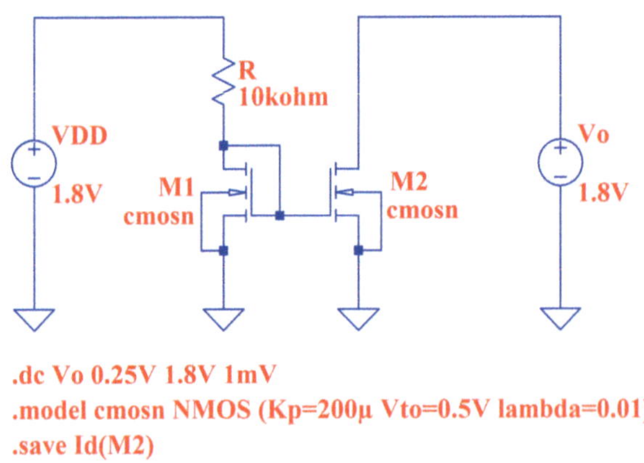

Fig. 1.19 Basic MOS current mirror with limited *DC sweep analysis* for output voltage V_O

save Id(M2)" is introduced in the file, with Id(M2) representing the output current of the circuit I_O.

The simulated value of the output current I_O is approximately 98.4 μA, very close to the theoretical value of 98.6 μA, expressed by Eq. (1.51).

The output current I_O is linearly increasing with the output voltage, but only for an output voltage superior to or equal to the minimal output voltage V_{Omin}, expressed by Eq. (1.53).

For a more facile evaluation with LTspice of the output resistance of the basic MOS current mirror from Fig. 1.17, the range of variation for the output voltage V_O is limited (a *DC sweep analysis* is performed in LTspice, with the variation range for the output voltage V_O being set between 0.25 and 1.8 V). The circuit used for this simulation is presented in Fig. 1.19, with the graphical representation of the output current I_O being shown in Fig. 1.20. In order to directly view the result of the

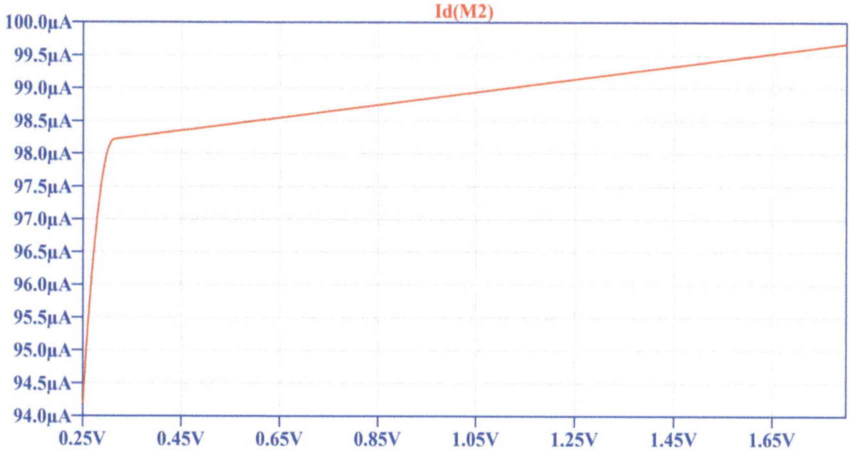

Fig. 1.20 Graphical representation of the characteristic $I_O(V_O)$, for a limited range of the output voltage V_O

simulation, the directive ".save Id(M2)" is introduced in the file, with Id (M2) representing the output current of the circuit I_O.

In this simulation, the slope of the characteristic $I_O(V_O)$ is equal to $1/R_O$, with R_O being the output resistance of the MOS current source from Fig. 1.19. In consequence, the output resistance will have the following value:

$$R_O = \frac{1.8\,\text{V} - 0.8\,\text{V}}{99.675\,\mu\text{A} - 98.694\,\mu\text{A}} = \frac{1\,\text{V}}{0.981\,\mu\text{A}} = 1.019\,\text{M}\Omega, \tag{1.54}$$

very close to the value of the theoretical output resistance, computed in Eq. (1.52).

The simulated value of the minimal output voltage is equal to 0.313 V, very close to the theoretical value of V_{Omin}, expressed by Eq. (1.53).

In order to graphically represent the dependence of the output current I_O on the output voltage variations, for multiple values of channel-length parameter λ_n (0.01, 0.02, 0.03, 0.04 and 0.05 V^{-1}), a *DC sweep analysis* is performed in LTspice for a range of the output voltage V_O between 0.7 and 1.8 V and also a parametrical simulation for $x = \lambda_n$, between 0.01 and 0.05 V^{-1}, with a step of 0.01 V^{-1}. The circuit used for this simulation is presented in Fig. 1.21.

The parametrical simulation results for the output characteristics of the current mirror $I_O(V_O)$ are illustrated in Fig. 1.22. In order to directly view the results of the simulation, the directive ".save Id(M2)" is introduced in the file, with Id (M2) representing the output current of the circuit I_O.

In Fig. 1.22, the line with the greatest slope represents the simulation for the largest value of channel-length modulation parameter $\lambda_n = 0.05\,\text{V}^{-1}$, equivalent to the shortest channel length of transistor M2 and to the smallest value of the output resistance R_O for the basic MOS current mirror from Fig. 1.21.

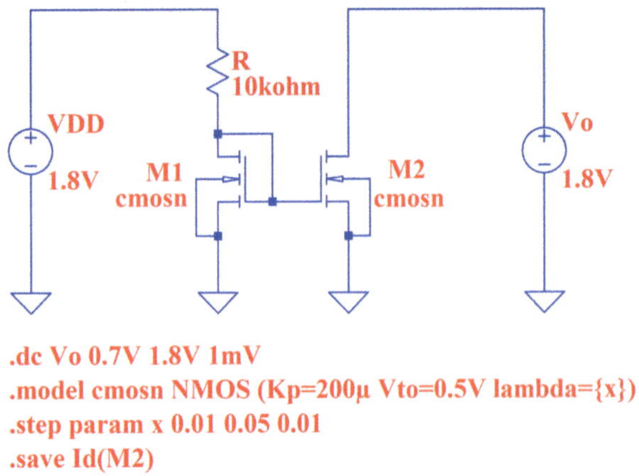

Fig. 1.21 Circuit used for obtaining the graphical representation of the characteristics $I_O(V_O)$, for multiple values of the channel-length parameter λ_n

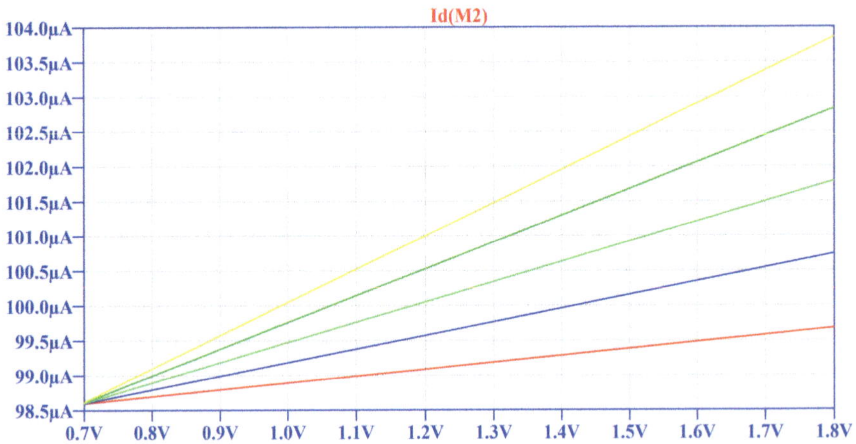

Fig. 1.22 Graphical representation of the characteristics $I_O(V_O)$, for multiple values of channel-length parameter λ_n

The line with the smallest slope represents the simulation for the smallest value of channel-length modulation parameter $\lambda_n = 0.01\,\mathrm{V}^{-1}$, equivalent to the largest channel length of transistor M2 and to the largest value of the output resistance R_O for the basic MOS current mirror from Fig. 1.21.

In order to evaluate the dependence of the output current I_O on the supply voltage V_{DD} (Eqs. (1.39) and (1.43)), a *DC sweep analysis* is performed in LTspice in Fig. 1.23, for a variation of the supply voltage V_{DD} between 0 and 1.8 V.

The simulation result for the characteristic $I_O(V_{DD})$ of the current mirror is illustrated in Fig. 1.24. In order to directly view the result of the simulation, the directive ".save Id(M2)" is introduced in the file, with Id(M2) representing the output current of the circuit I_O.

Fig. 1.23 Circuit used for obtaining the graphical representation of the dependence $I_O(V_{DD})$

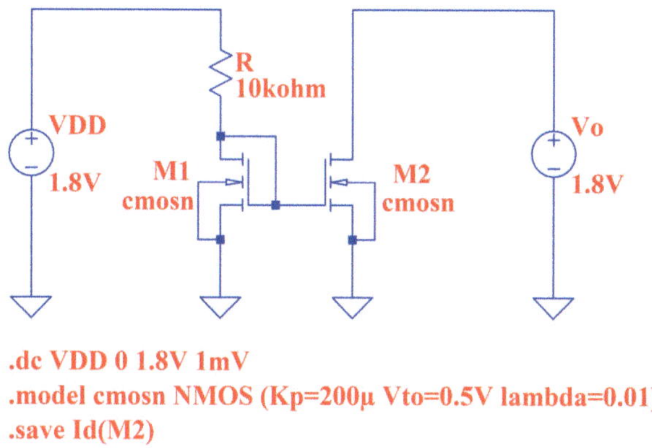

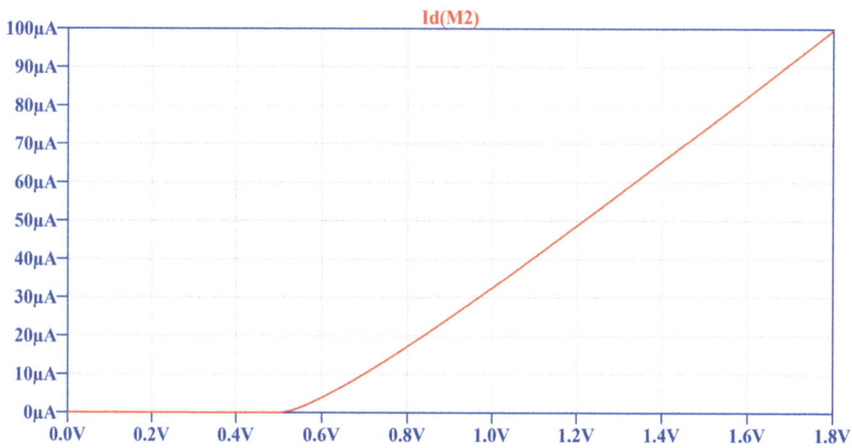

Fig. 1.24 Graphical representation of the characteristic $I_O(V_{DD})$

For a value of the supply voltage superior to a minimal value, the output current strongly increases when the supply voltage increases.

Because the mirroring factor I_O/I_{REF} is affected by the channel-length modulation effect (Eq. (1.42)), Fig. 1.25 shows the simulated dependence of this mirroring factor on the output voltage V_O.

So, the basic MOS current mirror introduces relatively large mirroring errors, especially for large values of the output voltage V_O.

1.3.4 Basic MOS Current Mirror with Multiple Outputs

The basic MOS current mirror from Fig. 1.16 can be extended for generating multiple output currents, having, in the general case, different values. In Fig. 1.26

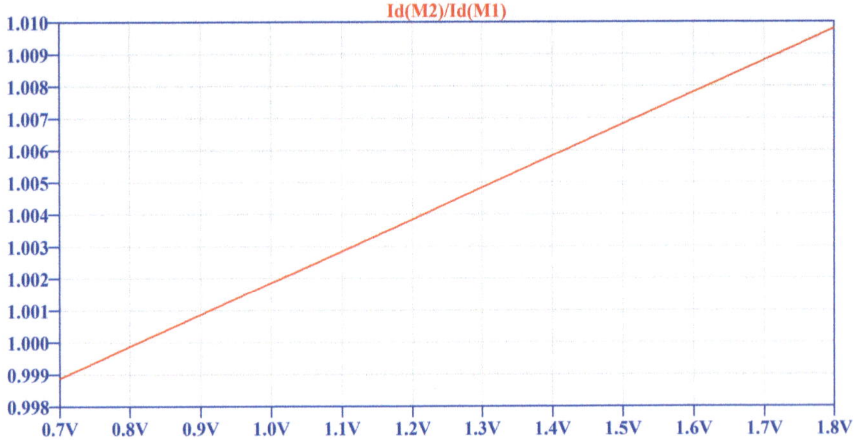

Fig. 1.25 The mirroring factor I_O/I_{REF} as a function of the output voltage V_O

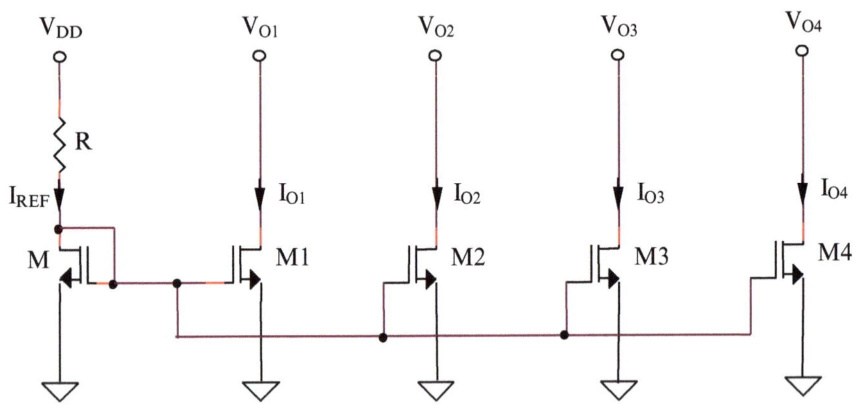

Fig. 1.26 Basic MOS current mirror with four output currents

is shown the basic MOS current mirror with four output currents: I_{O1}, I_{O2}, I_{O3} and I_{O4}.

1.3.4.1 Expressions of the Output Currents, I_{O1}, I_{O2}, I_{O3} and I_{O4}

In order to obtain different values of these output currents, different aspect ratios for transistors M1, M2, M3 and M4 must be set. Using a similar analysis with that made for basic MOS current mirror from Fig. 1.16, the reference current of the basic MOS current mirror with four outputs from Fig. 1.26, I_{REF}, will have the same expression (1.39). The approximated expressions of the output currents will be derived from Eq. (1.43):

$$I_{O1} \cong I_{REF} \frac{\left(\frac{W}{L}\right)_1}{\left(\frac{W}{L}\right)}, \tag{1.55}$$

$$I_{O2} \cong I_{REF} \frac{\left(\frac{W}{L}\right)_2}{\left(\frac{W}{L}\right)}, \tag{1.56}$$

$$I_{O3} \cong I_{REF} \frac{\left(\frac{W}{L}\right)_3}{\left(\frac{W}{L}\right)} \tag{1.57}$$

and

$$I_{O4} \cong I_{REF} \frac{\left(\frac{W}{L}\right)_4}{\left(\frac{W}{L}\right)}, \tag{1.58}$$

where $\left(\frac{W}{L}\right)_1$, $\left(\frac{W}{L}\right)_2$, $\left(\frac{W}{L}\right)_3$, $\left(\frac{W}{L}\right)_4$ and $\left(\frac{W}{L}\right)$ represent the aspect ratios of transistors M1, M2, M3, M4 and M, respectively.

1.3.4.2 Design Example

In order to confirm the previously presented theoretical estimated results, the basic MOS current mirror from Fig. 1.26 is simulated with LTspice using the following values: $K_n' = 200\,\mu\text{A/V}^2$, $\lambda_n = 0.01\,\text{V}^{-1}$, $V_{Tn} = 0.5\,\text{V}$ and $V_{DD} = 1.8\,\text{V}$. MOS transistors have the following aspect ratios: $\left(\frac{W}{L}\right)_1 = \frac{2\,\mu\text{m}}{0.5\,\mu\text{m}} = 4$, $\left(\frac{W}{L}\right)_2 = \frac{4\,\mu\text{m}}{0.5\,\mu\text{m}} = 8$, $\left(\frac{W}{L}\right)_3 = \frac{6\,\mu\text{m}}{0.5\,\mu\text{m}} = 12$, $\left(\frac{W}{L}\right)_4 = \frac{8\,\mu\text{m}}{0.5\,\mu\text{m}} = 16$ and $\left(\frac{W}{L}\right) = \frac{2\,\mu\text{m}}{0.5\,\mu\text{m}} = 4$, respectively, while the value of the resistor is $R = 10\,\text{k}\Omega$. The circuit used for illustrating the output characteristics of the MOS current mirror with four outputs, $I_{O1}(V_{O1})$, $I_{O2}(V_{O2})$, $I_{O3}(V_{O3})$ and $I_{O4}(V_{O4})$, respectively, is shown in Fig. 1.27. In order to sweep all four

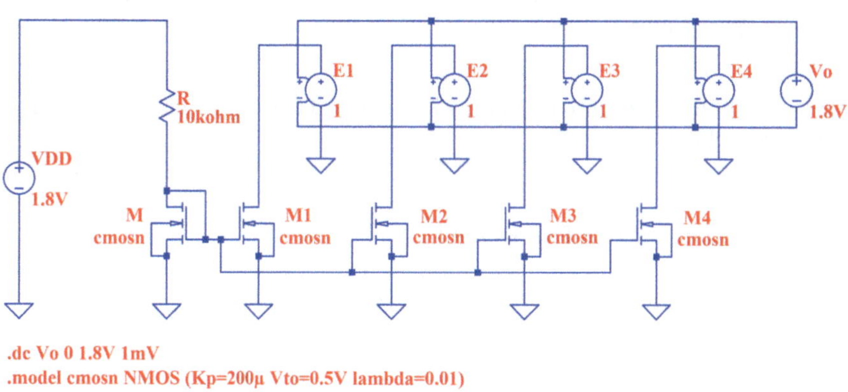

Fig. 1.27 Basic MOS current mirror with four outputs (*DC sweep analysis* for output voltage V_O)

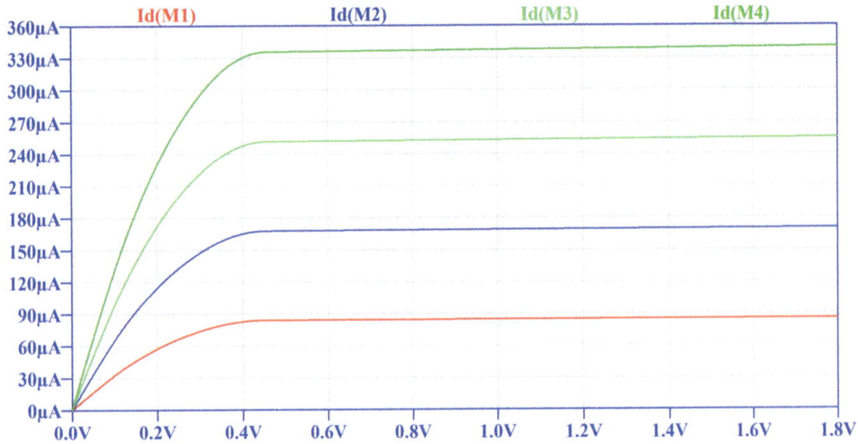

Fig. 1.28 Output characteristics of the basic MOS current mirror with four outputs from Fig. 1.27

output voltages of the multiple current mirror synchronously, four voltage-controlled voltage sources are used in LTspice ($E1$, $E2$, $E3$ and $E4$, respectively), each one having the same controlling voltage V_O and the same gain factor, equal to 1. In consequence, sweeping the voltage V_O between 0 and 1.8 V will sweep similarly $E1$, $E2$, $E3$ and $E4$ voltages between 0 and 1.8 V. Because these four voltages represent the output voltages for each output of the current mirror, the sweeping of the voltage V_O will generate four output characteristics of the multiple basic MOS current mirror from Fig. 1.27 (shown in Fig. 1.28). It is necessary to set four current markers in LTspice, in the drain terminals of transistors M1–M4. The simulated values of the output currents, I_{O1}, I_{O2}, I_{O3} and I_{O4}, confirm the theoretical estimations (Eqs. (1.55)–(1.58)).

1.3.5 *MOS Current Source with Increased Output Resistance (Version I)*

The circuit of the MOS current source with increased output resistance (version I) is illustrated in Fig. 1.29.

1.3.5.1 Output Current Expression I_O

The expression of the supply voltage can be written as follows:

$$V_{DD} = R_1 I_{REF} + V_{GS1}. \tag{1.59}$$

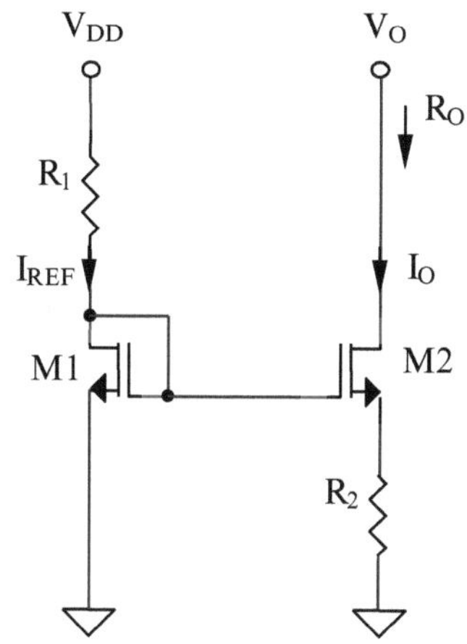

Fig. 1.29 MOS current source with increased output resistance (version I)

Because of the biasing in saturation of transistor M1, the reference current can be expressed as

$$I_{REF} = \frac{K'_n}{2}\left(\frac{W}{L}\right)_1 (V_{GS1} - V_T)^2. \tag{1.60}$$

In order to determine the expression of I_{REF} current, the system of equations containing previous two relations must be solved. Because we have the $V_{GS1} \geq V_T$ criteria available for choosing the correct solution of the second-order equation that will be obtained by analysing these relations (because of the biasing in saturation region of transistor M1), the expression of I_{REF} current from Eq. (1.60) must be replaced in Eq. (1.59), resulting in the following second-order equation:

$$V_{DD} = R_1 \frac{K'_n}{2}\left(\frac{W}{L}\right)_1 (V_{GS1} - V_T)^2 + V_{GS1}. \tag{1.61}$$

The solutions of this equation are

$$(V_{GS1})_A = V_T + \frac{1}{K'_n\left(\frac{W}{L}\right)_1 R_1}\left(\sqrt{1 + 2K'_n\left(\frac{W}{L}\right)_1 R_1 (V_{DD} - V_T)} - 1\right) \tag{1.62}$$

and

$$(V_{GS1})_B = V_T - \frac{1}{K'_n\left(\frac{W}{L}\right)_1 R_1}\left(\sqrt{1 + 2K'_n\left(\frac{W}{L}\right)_1 R_1 (V_{DD} - V_T)} + 1\right). \tag{1.63}$$

Because only $(V_{GS1})_A \geq V_T$, it will represent the mathematical solution of Eq. (1.61), resulting in the following expression of the reference current:

$$I_{REF} = \frac{1}{K_n'\left(\frac{W}{L}\right)_1 R_1^2}$$
$$\times \left[K_n'\left(\frac{W}{L}\right)_1 R_1(V_{DD} - V_T) - \sqrt{1 + 2K_n'\left(\frac{W}{L}\right)_1 R_1(V_{DD} - V_T) + 1}\right]. \quad (1.64)$$

In order to determine the expression of the output current I_O, the following relations can be written:

$$(V_{GS1})_A = V_{GS2} + I_O R_2 \quad (1.65)$$

and

$$I_O = \frac{K_n'}{2}\left(\frac{W}{L}\right)_2 (V_{GS2} - V_T)^2. \quad (1.66)$$

Replacing I_O from Eq. (1.66) in Eq. (1.65) and using the value of $(V_{GS1})_A$ given by Eq. (1.62) result in the following second-order equation, having as unknown value V_{GS2} voltage:

$$(V_{GS1})_A = V_{GS2} + \frac{K_n'}{2}\left(\frac{W}{L}\right)_2 R_2(V_{GS2} - V_T)^2. \quad (1.67)$$

The solution of the previous equation corresponds to the biasing in saturation region of transistor M2 ($V_{GS2} > V_T$). Replacing this value of V_{GS2} voltage in Eq. (1.66), it is possible to obtain the value of the output current I_O.

1.3.5.2 Output Resistance Expression R_O

Considering the expression of the dynamic resistance viewed in the drain terminal of the MOS transistor (determined using Eq. (A.10) from Annex), the output resistance of the current source from Fig. 1.29 can be expressed as follows:

$$R_O = r_{ds2}(1 + g_{m2}R_2) = \frac{1}{\lambda_n I_{D2}}(1 + g_{m2}R_2) = \frac{1}{\lambda_n I_O}(1 + g_{m2}R_2). \quad (1.68)$$

1.3.5.3 Minimal Output Voltage V_{Omin}

In order to bias transistor M2 in saturation, the fundamental condition $V_{DS2} \geq V_{GS2} - V_T$ must be fulfilled. It results in the following expression of the minimal output voltage of the current source from Fig. 1.29:

$$V_{O\,min} = V_{GS2} - V_T + I_O R_2. \tag{1.69}$$

By neglecting the channel-length modulation effect that affects the operation of transistor M2, the squaring characteristic of this transistor can be expressed as

$$I_O \cong \frac{K'_n}{2}\left(\frac{W}{L}\right)(V_{GS2} - V_T)^2. \tag{1.70}$$

From the previous two relations, another equivalent mathematical expression of the minimal output voltage results:

$$V_{O\,min} \cong \sqrt{\frac{2I_O}{K'_n\left(\frac{W}{L}\right)}} + I_O R_2. \tag{1.71}$$

1.3.5.4 Design Examples

In order to confirm the previously presented theoretical estimated results, the MOS current source from Fig. 1.29 is simulated with LTspice using the following values: $K'_n = 200\,\mu\text{A/V}^2$, $\lambda_n = 0.01\,\text{V}^{-1}$, $V_{Tn} = 0.5\,\text{V}$ and $V_{DD} = 1.8\,\text{V}$. MOS transistors have the aspect ratios $\left(\frac{W}{L}\right)_1 = \left(\frac{W}{L}\right)_2 = \frac{5\,\mu\text{m}}{0.5\,\mu\text{m}} = 10$, while the values of resistors are $R_1 = 10\,\text{k}\Omega$ and $R_2 = 1\,\text{k}\Omega$.

The theoretical value of the gate-source voltage of transistor M1 for the current source from Fig. 1.29 could be determined using Eq. (1.61):

$$\frac{200\,\mu\text{A/V}^2}{2} \times 10 \times 10\text{k}\Omega \times (V_{GS1} - 0.5\,\text{V})^2 + V_{GS1} = 1.8\,\text{V}, \tag{1.72}$$

resulting in the value $V_{GS1} = 0.814\,\text{V}$, corresponding to a biasing in saturation region of transistor M1.

The theoretical value of the gate-source voltage of transistor M2 for the current source from Fig. 1.29 could be determined using Eq. (1.67):

$$\frac{200\,\mu\text{A/V}^2}{2} \times 10 \times 1\,\text{k}\Omega \times (V_{GS2} - 0.5\text{V})^2 + V_{GS2} = 0.814\text{V}, \tag{1.73}$$

resulting in the value $V_{GS2} = 0.751\,\text{V}$, corresponding to a biasing in saturation region of transistor M2. Replacing this value in Eq. (1.66) results in

$$I_O \cong \frac{200\,\mu\text{A/V}^2}{2} \times 10 \times (0.751\,\text{V} - 0.5\,\text{V})^2 = 63\,\mu\text{A}. \tag{1.74}$$

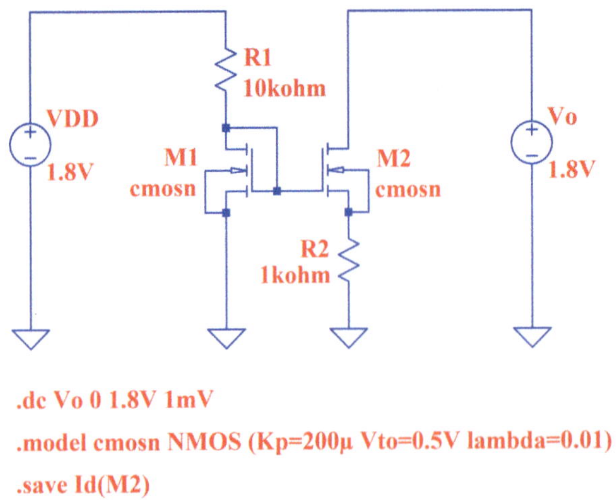

Fig. 1.30 MOS current source with *DC sweep analysis* for output voltage V_O

In order to determine the theoretical value of the output resistance, the transconductance g_{m2} of transistor M2 must be determined:

$$g_{m2} = \sqrt{2 \times 200\,\mu\text{A}/\text{V}^2 \times 10 \times 63\,\mu\text{A}} = 502\,\mu\text{A}/\text{V}. \tag{1.75}$$

The theoretical value of the output resistance can be obtained using Eq. (1.68):

$$R_O = \frac{1}{0.01\,\text{V}^{-1} \times 63\,\mu\text{A}}(1 + 502\,\mu\text{A}/\text{V} \times 1\,\text{k}\Omega) = 2.384\,\text{M}\Omega. \tag{1.76}$$

The theoretical value of the minimal output voltage can be obtained using Eq. (1.69):

$$V_{O\,min} = 0.751\,\text{V} - 0.5\,\text{V} + 63\,\mu\text{A} \times 1\,\text{k}\Omega = 0.314\,\text{V}. \tag{1.77}$$

The circuit used for simulating the output characteristic $I_O(V_O)$ of the current source is shown in Fig. 1.30.

Setting a variation of the output voltage V_O between 0 and 1.8 V (a *DC sweep analysis* in LTspice), the graphical representation of the output current I_O is shown in Fig. 1.31 (the output characteristic of the MOS current source from Fig. 1.30). In order to directly view the result of the simulation, the directive ".save Id(M2)" is introduced in the file, with Id(M2) representing the output current of the circuit I_O.

The simulated value of the output current I_O is approximately 62.8 µA, very close to the theoretical value of 63 µA, given by Eq. (1.74).

The output current I_O is linearly increasing with the output voltage, but only for an output voltage superior to or equal to the minimal output voltage V_{Omin}, expressed by Eq. (1.77).

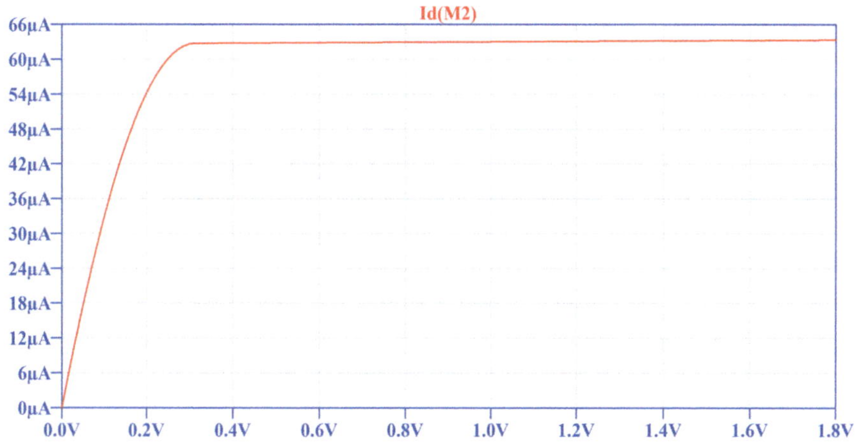

Fig. 1.31 Output characteristic $I_O(V_O)$ for the current source presented in Fig. 1.30

Fig. 1.32 MOS current source with limited range of the *DC sweep analysis* for output voltage variation

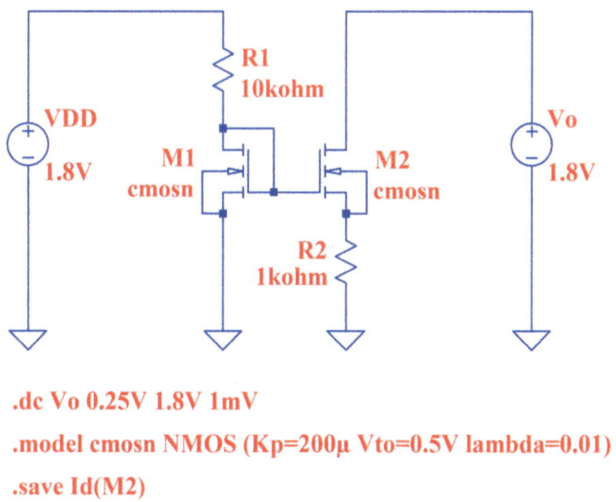

For a more facile evaluation with LTspice of the output resistance of the MOS current source from Fig. 1.30, the range of variation for the output voltage V_O is limited (*DC sweep analysis* in LTspice, with the variation range for the output voltage V_O being set between 0.25 and 1.8 V). The circuit used for this simulation is presented in Fig. 1.32, with the graphical representation of the output current I_O being shown in Fig. 1.33. In order to directly view the result of the simulation, the directive ".save Id(M2)" is introduced in the file, with Id(M2) representing the output current of the circuit I_O.

In this simulation, the slope of the characteristic $I_O(V_O)$ is equal to $1/R_O$, with R_O being the output resistance of the MOS current source from Fig. 1.32. In consequence, the output resistance will have the following value:

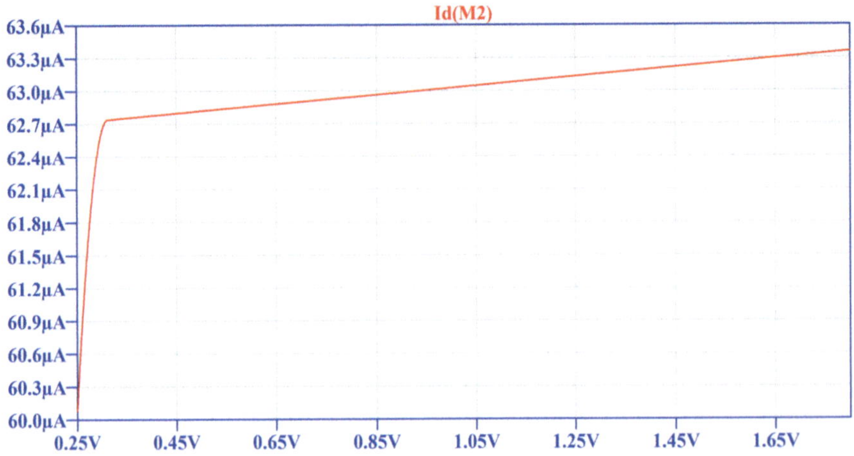

Fig. 1.33 Graphical representation of the characteristic $I_O(V_O)$, for a limited range of the output voltage V_O

$$R_O = \frac{1.8\,\text{V} - 0.8\,\text{V}}{63.358\,\mu\text{A} - 62.941\,\mu\text{A}} = \frac{1\,\text{V}}{0.417\,\mu\text{A}} = 2.398\,\text{M}\Omega, \tag{1.78}$$

very close to the value of the theoretical output resistance, computed in Eq. (1.76).

The simulated value of the minimal output voltage is equal to 0.313 V, very close to the theoretical value of V_{Omin}, given by Eq. (1.77).

In order to graphically represent the dependence of the output current I_O on the output voltage variation, for multiple values of channel-length parameter λ_n (0.01, 0.02, 0.03, 0.04 and 0.05 V^{-1}), a *DC sweep analysis* is performed in LTspice for a range of the output voltage V_O between 0.9 and 1.8 V and also a parametrical simulation for $x = \lambda_n$, between 0.01 and 0.05 V^{-1}, with a step of 0.01 V^{-1}. The circuit used for this simulation is presented in Fig. 1.34.

The parametrical simulation results for the output characteristics of the current source $I_O(V_O)$ are illustrated in Fig. 1.35. In order to directly view the results of the simulation, the directive ".save Id(M2)" is introduced in the file, with Id (M2) representing the output current of the circuit I_O.

In Fig. 1.35, the line with the greatest slope represents the simulation for the largest value of channel-length modulation parameter $\lambda_n = 0.05\,\text{V}^{-1}$, equivalent to the shortest channel length of transistor M2 and to the smallest value of the output resistance R_O for the MOS current source from Fig. 1.34.

The line with the smallest slope represents the simulation for the smallest value of channel-length modulation parameter $\lambda_n = 0.01\,\text{V}^{-1}$, equivalent to the largest channel length of transistor M2 and to the largest value of the output resistance R_O for the MOS current source from Fig. 1.34.

In order to evaluate the dependence of the output current I_O on the supply voltage V_{DD}, a *DC sweep analysis* is performed in LTspice in Fig. 1.36, for a variation of the supply voltage V_{DD} between 0 and 1.8 V.

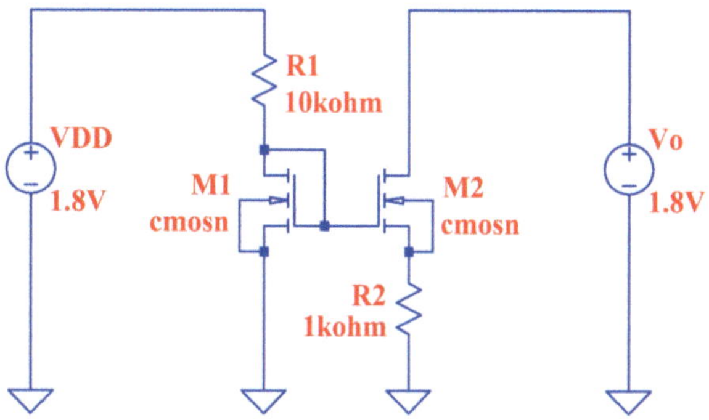


```
.dc Vo 0.9V 1.8V 1mV
.model cmosn NMOS (Kp=200μ Vto=0.5V lambda={x})
.step param x 0.01 0.05 0.01
.save Id(M2)
```

Fig. 1.34 Circuit used for obtaining the graphical representation of the characteristics $I_O(V_O)$, for multiple values of channel-length parameter λ_n

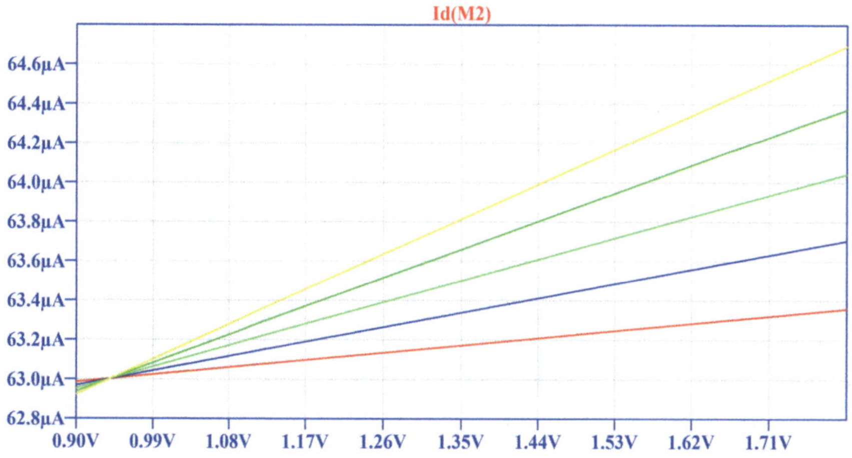

Fig. 1.35 Graphical representation of the characteristics $I_O(V_O)$, for multiple values of channel-length parameter λ_n

The simulation results for the characteristic of the current source $I_O(V_{DD})$ are illustrated in Fig. 1.37. In order to directly view the result of the simulation, the directive ".save Id(M2)" is introduced in the file, with Id(M2) representing the output current of the circuit I_O.

For a value of the supply voltage superior to a minimal value, the output current strongly increases when the supply voltage increases.

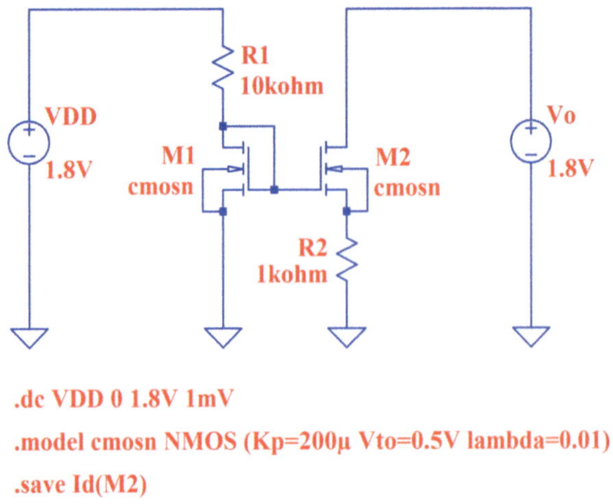

Fig. 1.36 Circuit used for obtaining the graphical representation of the dependence $I_O(V_{DD})$

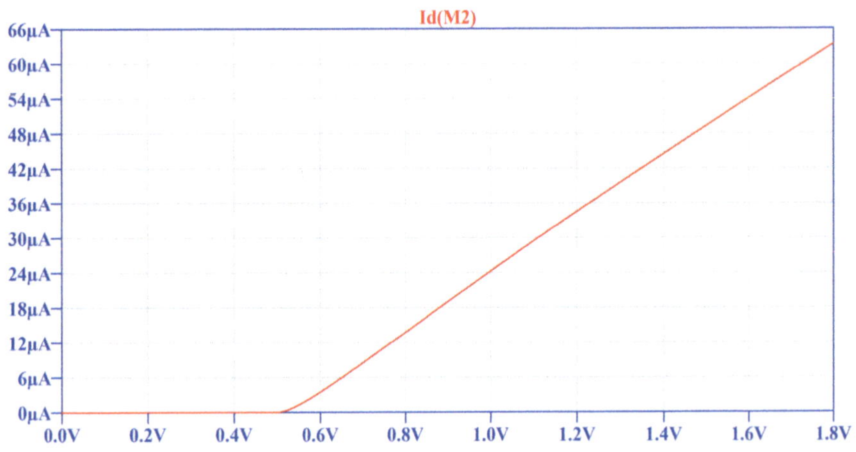

Fig. 1.37 Graphical representation of the characteristic $I_O(V_{DD})$

1.3.6 MOS Current Source with Increased Output Resistance (Version II)

The circuit of the MOS current source with increased output resistance (version II) is illustrated in Fig. 1.38.

1.3.6.1 Output Current Expression I_O

The expression of the supply voltage can be written as follows:

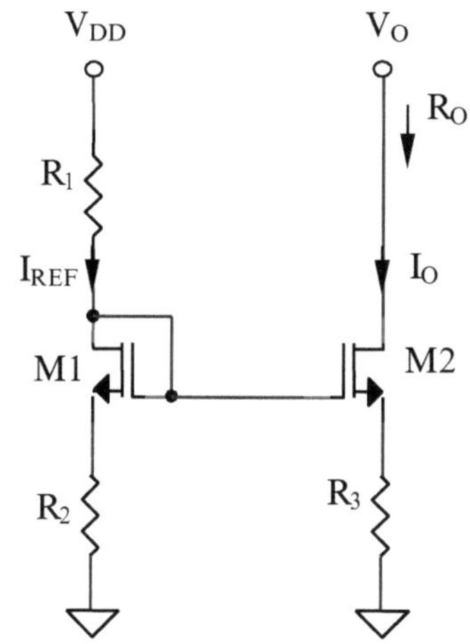

Fig. 1.38 MOS current source with increased output resistance (version II)

$$V_{DD} = (R_1 + R_2)I_{REF} + V_{GS1}. \tag{1.79}$$

Because of the biasing in saturation of the transistor M1, the reference current can be expressed as

$$I_{REF} = \frac{K_n'}{2}\left(\frac{W}{L}\right)_1 (V_{GS1} - V_T)^2. \tag{1.80}$$

In order to determine the expression of the current I_{REF}, the system of equations containing previous two relations must be solved. Because we have the $V_{GS1} \geq V_T$ criteria available for choosing the correct solution of the second-order equation that will be obtained by analysing these relations (because of the biasing in saturation region of transistor M1), the expression of I_{REF} current from Eq. (1.80) must be replaced in Eq. (1.79), resulting in the following second-order equation:

$$V_{DD} = (R_1 + R_2)\frac{K_n'}{2}\left(\frac{W}{L}\right)_1 (V_{GS1} - V_T)^2 + V_{GS1}. \tag{1.81}$$

The solutions of this equation are

$$(V_{GS1})_A = V_T + \frac{1}{K_n'\left(\frac{W}{L}\right)_1 (R_1 + R_2)} \times \left(\sqrt{1 + 2K_n'\left(\frac{W}{L}\right)_1 (R_1 + R_2)(V_{DD} - V_T)} - 1\right) \tag{1.82}$$

Fig. 1.39 MOS current source with *DC sweep analysis* for output voltage V_O

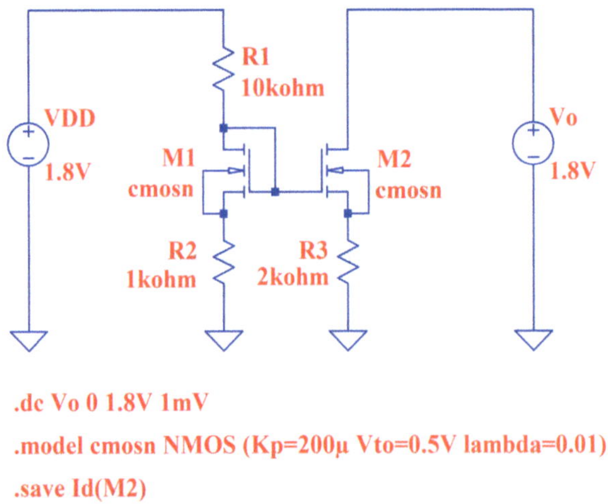

$$\begin{aligned}&\frac{200\,\mu\text{A}/\text{V}^2}{2}\times 10\times 1\,\text{k}\Omega\times(0.801\,\text{V}-0.5\,\text{V})^2+0.801\,\text{V}\\&=\frac{200\,\mu\text{A}/\text{V}^2}{2}\times 10\times 2\,\text{k}\Omega\times(V_{GS2}-0.5\,\text{V})^2+V_{GS2},\end{aligned}\tag{1.93}$$

resulting in the value $V_{GS2}=0.759\,\text{V}$, corresponding to a biasing in saturation region of transistor M2. Replacing this value in Eq. (1.86) results in

$$I_O\cong\frac{200\,\mu\text{A}/\text{V}^2}{2}\times 10\times(0.759\,\text{V}-0.5\,\text{V})^2=67.08\,\mu\text{A}.\tag{1.94}$$

In order to determine the theoretical value of the output resistance, the transconductance g_{m2} of transistor M2 must be determined:

$$g_{m2}=\sqrt{2\times 200\,\mu\text{A}/\text{V}^2\times 10\times 67.08\,\mu\text{A}}=518\,\mu\text{A}/\text{V}.\tag{1.95}$$

The theoretical value of the output resistance can be obtained using Eq. (1.88):

$$R_O=\frac{1}{0.01\,\text{V}^{-1}\times 67.08\,\mu\text{A}}(1+502\,\mu\text{A}/\text{V}\times 2\,\text{k}\Omega)=2.99\,\text{M}\Omega.\tag{1.96}$$

The theoretical value of the minimal output voltage can be obtained using Eq. (1.89):

$$V_{O\,min}=0.759\,\text{V}-0.5\,\text{V}+67.08\,\mu\text{A}\times 2\,\text{k}\Omega=0.393\,\text{V}.\tag{1.97}$$

The circuit used for simulating the output characteristic $I_O(V_O)$ of the current source is shown in Fig. 1.39.

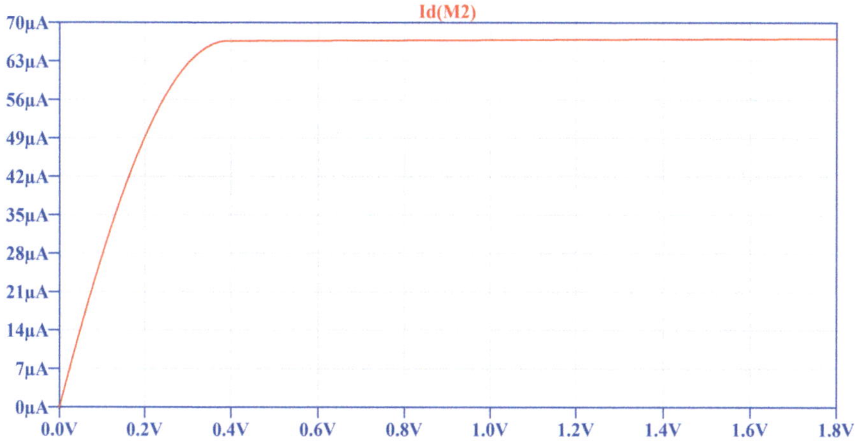

Fig. 1.40 Output characteristic $I_O(V_O)$ for the current source presented in Fig. 1.39

Setting a variation of the output voltage V_O between 0 and 1.8 V (a *DC sweep analysis* in LTspice), the graphical representation of the output current I_O is shown in Fig. 1.40 (the output characteristic of the MOS current source from Fig. 1.39). In order to directly view the result of the simulation, the directive ".save Id(M2)" is introduced in the file, with Id(M2) representing the output current of the circuit I_O.

The simulated value of the output current I_O is approximately 66.94 µA, very close to the theoretical value of 67.08 µA, expressed by Eq. (1.94).

The output current I_O is linearly increasing with the output voltage, but only for an output voltage superior to or equal to the minimal output voltage V_{Omin}, expressed by Eq. (1.97).

For a more facile evaluation with LTspice of the output resistance of the MOS current source from Fig. 1.39, the range of variation for the output voltage V_O is limited (*DC sweep analysis* in LTspice, with the variation range for the output voltage V_O being set between 0.3 and 1.8 V). The circuit used for this simulation is presented in Fig. 1.41, with the graphical representation of the output current I_O being shown in Fig. 1.42. In order to directly view the result of the simulation, the directive ".save Id(M2)" is introduced in the file, with Id(M2) representing the output current of the circuit I_O.

In this simulation, the slope of the characteristic $I_O(V_O)$ is equal to $1/R_O$, with R_O being the output resistance of the MOS current source from Fig. 1.41. In consequence, the output resistance will have the following value:

$$R_O = \frac{1.8\,\text{V} - 0.8\,\text{V}}{67.113\,\mu\text{A} - 66.782\,\mu\text{A}} = \frac{1\,\text{V}}{0.331\,\mu\text{A}} = 3.02\,\text{M}\Omega, \tag{1.98}$$

very close to the value of the theoretical output resistance, computed in Eq. (1.96).

The simulated value of the minimal output voltage is equal to 0.392 V, very close to the theoretical value of V_{Omin}, expressed by Eq. (1.97).

Fig. 1.41 MOS current source with limited range of the *DC sweep analysis* for output voltage variation

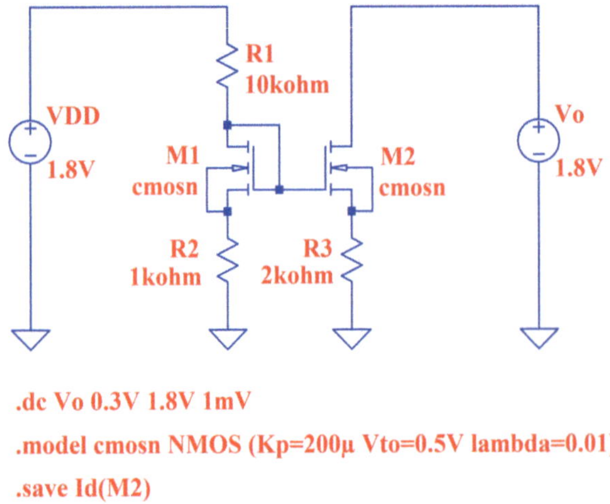

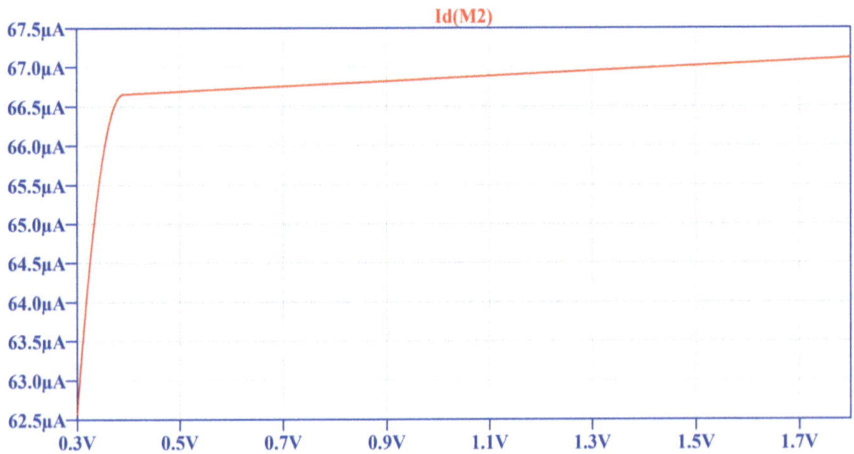

Fig. 1.42 Graphical representation of the characteristic $I_O(V_O)$, for a limited range of the output voltage V_O

In order to graphically represent the dependence of the output current I_O on the output voltage variation, for multiple values of the channel-length parameter λ_n (0.01, 0.02, 0.03, 0.04 and 0.05 V^{-1}), a *DC sweep analysis* is performed in LTspice for a range of the output voltage V_O between 0.9 and 1.8 V and also a parametrical simulation for $x = \lambda_n$, between 0.01 and 0.05 V^{-1}, with a step of 0.01 V^{-1}. The circuit used for this simulation is presented in Fig. 1.43.

The parametrical simulation results for the output characteristics of the current source $I_O(V_O)$ are illustrated in Fig. 1.44. In order to directly view the result of the simulation, the directive ".save Id(M2)" is introduced in the file, with Id (M2) representing the output current of the circuit I_O.

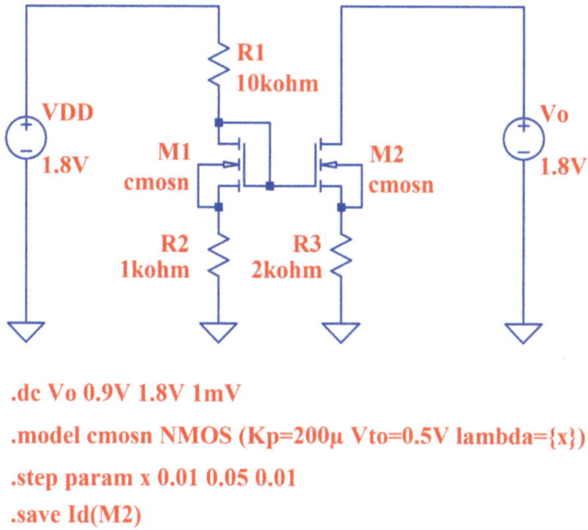

Fig. 1.43 Circuit used for obtaining the graphical representation of the characteristics $I_O(V_O)$, for multiple values of the channel-length parameter λ_n

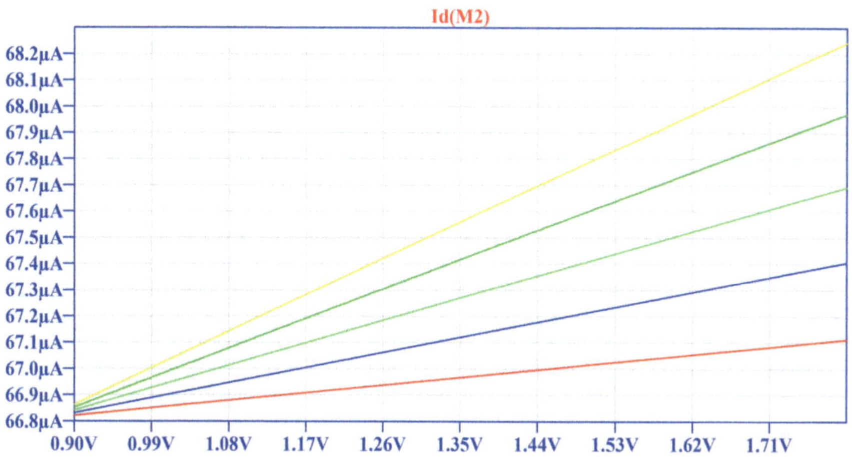

Fig. 1.44 Graphical representation of characteristics $I_O(V_O)$, for multiple values of channel-length parameter λ_n

In Fig. 1.44, the line with the greatest slope represents the simulation for the largest value of the channel-length modulation parameter $\lambda_n = 0.05\,\mathrm{V}^{-1}$, equivalent to the shortest channel length of transistor M2 and to the smallest value of the output resistance R_O for the MOS current source from Fig. 1.43.

The line with the smallest slope represents the simulation for the smallest value of the channel-length modulation parameter $\lambda_n = 0.01\,\mathrm{V}^{-1}$, equivalent to the largest channel length of transistor M2 and to the largest value of the output resistance R_O for the MOS current source from Fig. 1.43.

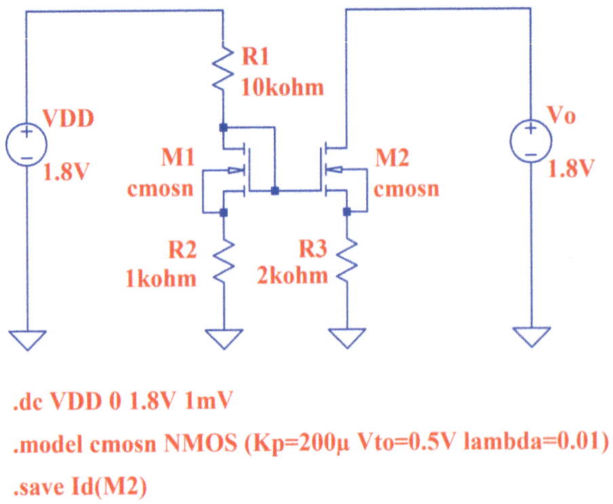

Fig. 1.45 Circuit used for obtaining the graphical representation of the dependence $I_O(V_{DD})$

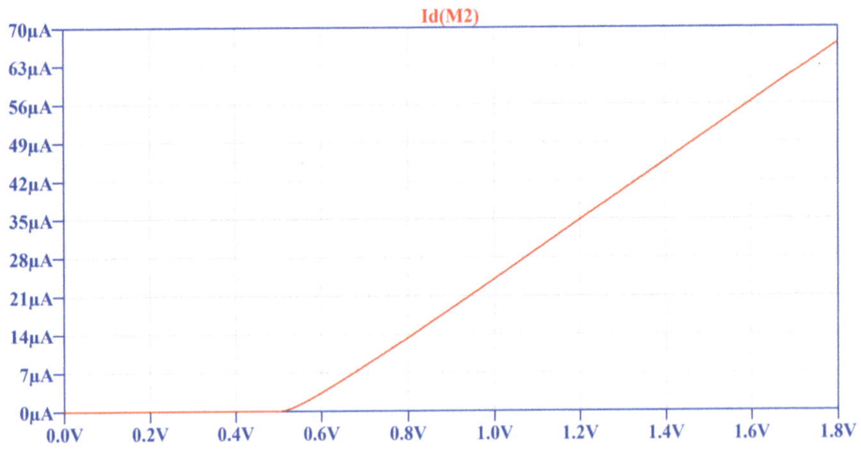

Fig. 1.46 Graphical representation of the characteristic $I_O(V_{DD})$

In order to evaluate the dependence of the output current I_O on the supply voltage V_{DD}, a *DC sweep analysis* is performed in LTspice in Fig. 1.45, for a variation of the supply voltage V_{DD} between 0 and 1.8 V.

The simulation result for the characteristic of the current source $I_O(V_{DD})$ is illustrated in Fig. 1.46. In order to directly view the result of the simulation, the directive ".save Id(M2)" is introduced in the file, with Id(M2) representing the output current of the circuit I_O.

For a value of the supply voltage superior to a minimal value, the output current strongly increases when the supply voltage increases.

1.4 Cascode CMOS Current Sources

In order to increase the output resistance R_O of a current source, the classical method is to use a cascode configuration.

1.4.1 Cascode MOS Current Source with Two Voltage Sources

A general architecture of an MOS cascode current source is represented by the cascode MOS current source with two voltage sources, as illustrated in Fig. 1.47.

1.4.1.1 Output Current Expression I_O

The output current I_O can be expressed as follows:

$$I_O \cong \frac{K_n'}{2}\left(\frac{W}{L}\right)_2 (V_{GS2} - V_T)^2 = \frac{K_n'}{2}\left(\frac{W}{L}\right)_2 (V_2 - V_T)^2. \tag{1.99}$$

1.4.1.2 Output Resistance Expression R_O

Considering the expression of the dynamic resistance viewed in the drain terminal of the MOS transistor (determined using Eq. (A.10) from Annex), the output resistance of the current source from Fig. 1.47 can be expressed as follows:

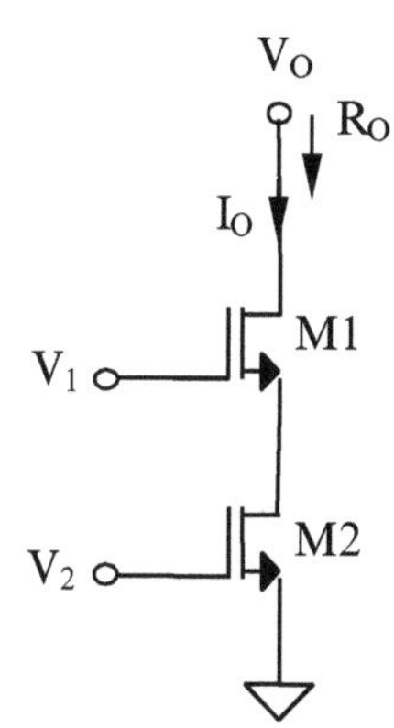

Fig. 1.47 Cascode MOS current source with two voltage sources

$$R_O = r_{ds1}(1 + g_{m1} r_{ds2}) \cong g_{m1} r_{ds1}^2. \tag{1.100}$$

1.4.1.3 Minimal Output Voltage V_{Omin}

In order to bias both transistors M1 and M2 in saturation, the fundamental condition $V_{DS} \geq V_{GS} - V_T$ must be fulfilled. Because V_{DS1} is equal to $V_O - (V_1 - V_{GS1})$, this condition for transistor M1 can be written as follows:

$$V_O - (V_1 - V_{GS1}) \geq V_{GS1} - V_T, \tag{1.101}$$

equivalent to

$$V_O \geq V_1 - V_T. \tag{1.102}$$

So, the expression of the minimal output voltage will be

$$V_{Omin} = V_1 - V_T. \tag{1.103}$$

The condition for biasing in saturation of transistor M2 will be

$$V_1 - V_{GS1} \geq V_{GS2} - V_T. \tag{1.104}$$

Because transistors M1 and M2 are identical and biased at the same drain current I_O, their gate-source voltage are equal, $V_{GS1} = V_{GS2} = V_2$. So, the previous relation becomes

$$V_1 \geq 2V_2 - V_T. \tag{1.105}$$

In order to minimize the minimal output voltage of the current source expressed by Eq. (1.103), the voltage V_1 must be set to be only a little bit superior to the value of $2V_2 - V_T$, with V_2 voltage chosen to be a little bit superior to its minimal value of V_T (for having transistor M2 biased in saturation region, it is necessary that $V_2 = V_{GS2} > V_T$).

1.4.1.4 Design Examples

In order to confirm the previously presented theoretical estimated results, the MOS current source from Fig. 1.47 is simulated with LTspice using the following values: $K'_n = 200\,\mu\text{A/V}^2$, $\lambda_n = 0.01\,\text{V}^{-1}$, $V_{Tn} = 0.5\,\text{V}$ and $V_{DD} = 1.8\,\text{V}$. MOS transistors have the aspect ratios $\left(\frac{W}{L}\right)_1 = \left(\frac{W}{L}\right)_2 = \frac{5\,\mu\text{m}}{0.5\,\mu\text{m}} = 10$, while the values of voltage sources are $V_1 = 0.7\,\text{V}$ and $V_2 = 0.6\,\text{V}$.

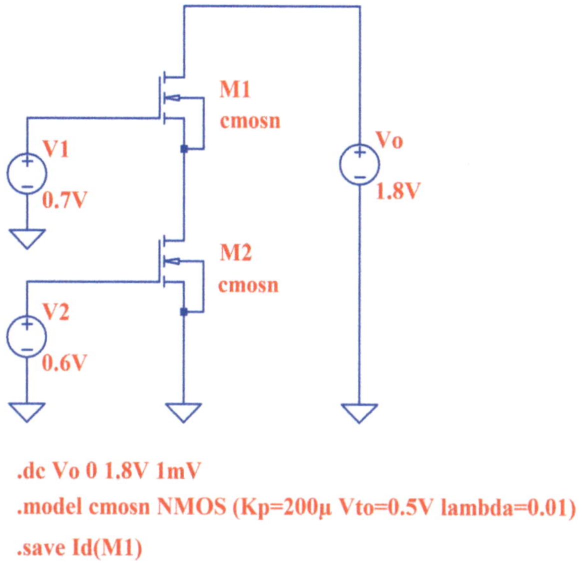

Fig. 1.48 MOS current source with *DC sweep analysis* for output voltage V_O

The theoretical value of the output current I_O can be determined using Eq. (1.99):

$$I_O \cong \frac{200\,\mu\text{A}/\text{V}^2}{2} \times 10 \times (0.6\,\text{V} - 0.5\,\text{V})^2 = 10\,\mu\text{A}. \tag{1.106}$$

In order to determine the theoretical value of the output resistance, the transconductance g_{m1} of transistor M1 must be determined:

$$g_{m1} = \sqrt{2 \times 200\,\mu\text{A}/\text{V}^2 \times 10 \times 10\,\mu\text{A}} = 200\,\mu\text{A}/\text{V}. \tag{1.107}$$

The theoretical value of the output resistance can be obtained using Eq. (1.100):

$$R_O = 200\,\mu\text{A}/\text{V} \times \left(\frac{1}{0.01\,\text{V}^{-1} \times 10\,\mu\text{A}}\right)^2 = 20\,\text{G}\Omega. \tag{1.108}$$

The theoretical value of the minimal output voltage can be obtained using Eq. (1.103):

$$V_{O\,min} = 0.7\,\text{V} - 0.5\,\text{V} = 0.2\,\text{V}. \tag{1.109}$$

The circuit used for simulating the output characteristic $I_O(V_O)$ of the current source with two voltage sources is shown in Fig. 1.48.

Setting a variation of the output voltage V_O between 0 and 1.8 V (a *DC sweep analysis* in LTspice), the graphical representation of the output current I_O is shown in Fig. 1.49 (the output characteristic of the MOS current source from Fig. 1.48).

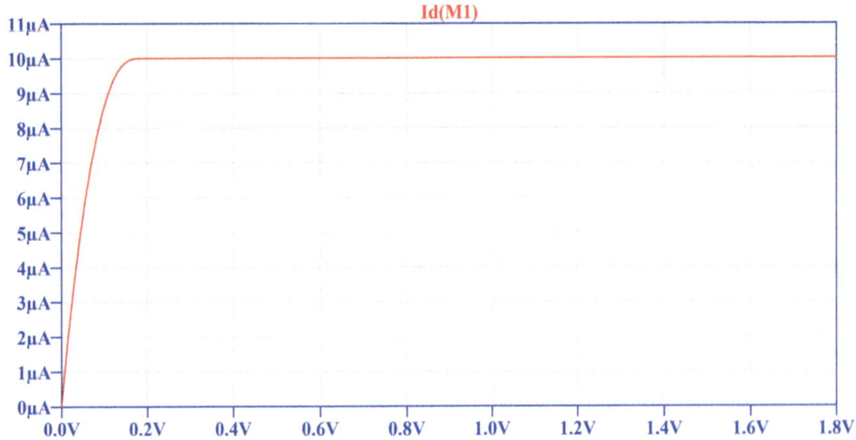

Fig. 1.49 Output characteristic $I_O(V_O)$ for the current source presented in Fig. 1.48

The simulated value of the output current I_O is approximately 10 µA, equal to the theoretical value, expressed by Eq. (1.106). In order to directly view the result of the simulation, the directive ".save Id(M1)" is introduced in the file, with Id (M1) representing the output current of the circuit I_O.

The output current I_O is linearly increasing with the output voltage, but only for an output voltage superior to or equal to the minimal output voltage V_{Omin}, expressed by Eq. (1.109).

For a more facile evaluation with LTspice of the output resistance of the MOS current source from Fig. 1.48, the range of variation for the output voltage V_O is limited (*DC sweep analysis* in LTspice, the variation range for the output voltage V_O being set between 0.15 and 1.8 V). The circuit used for this simulation is presented in Fig. 1.50, with the graphical representation of the output current I_O being shown in Fig. 1.51. In order to directly view the result of the simulation, the directive ".save Id (M1)" is introduced in the file, with Id(M1) representing the output current of the circuit I_O.

In this simulation, the slope of the characteristic $I_O(V_O)$ is equal to $1/R_O$, with R_O being the output resistance of the MOS current source from Fig. 1.50. In consequence, the output resistance will have the following value:

$$R_O = \frac{1\,\text{V}}{50.11\,\text{pA}} = 19.95\,\text{G}\Omega, \tag{1.110}$$

very close to the value of the theoretical output resistance, computed in Eq. (1.108).

The simulated value of the minimal output voltage is equal to 0.2 V, equal to the theoretical value of V_{Omin}, expressed by Eq. (1.109).

In order to graphically represent the dependence of the output current I_O on the output voltage variations for multiple values of channel-length parameter λ_n (0.01, 0.02, 0.03, 0.04 and 0.05 V^{-1}), a *DC sweep analysis* is performed in LTspice for a

Fig. 1.50 MOS current source with limited range of the *DC sweep analysis* for output voltage variation

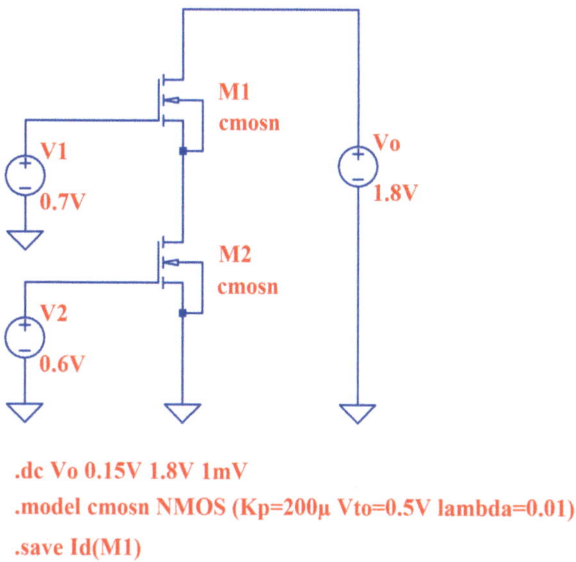

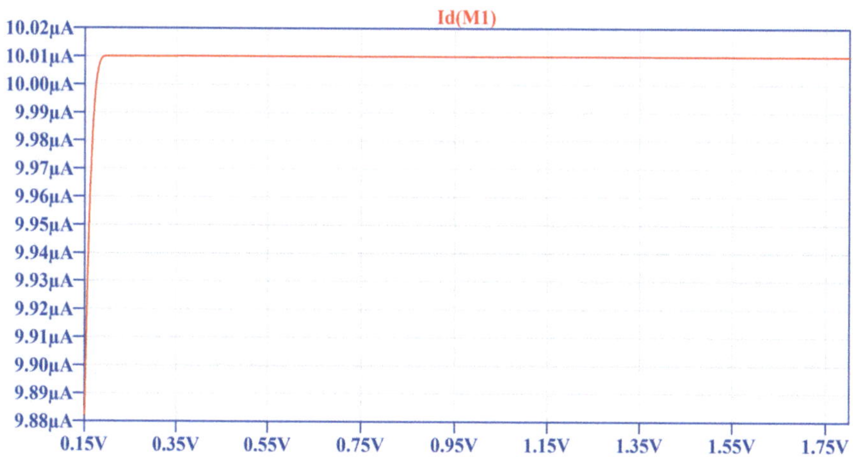

Fig. 1.51 Graphical representation of the characteristic $I_O(V_O)$, for a limited range of the output voltage V_O

variation range of the output voltage V_O between 0.4 and 1.8 V and also a parametrical simulation for $x = \lambda_n$, between 0.01 and 0.05 V^{-1}, with a step of 0.01 V^{-1}. The circuit used for this simulation is presented in Fig. 1.52.

The parametrical simulation results for the output characteristics of the current source $I_O(V_O)$ are illustrated in Fig. 1.53. In order to directly view the result of the simulation, the directive ".save Id(M1)" is introduced in the file, with Id (M1) representing the output current of the circuit I_O.

Fig. 1.52 Circuit used for obtaining the graphical representation of $I_O(V_O)$ characteristics for multiple values of the channel-length parameter λ_n

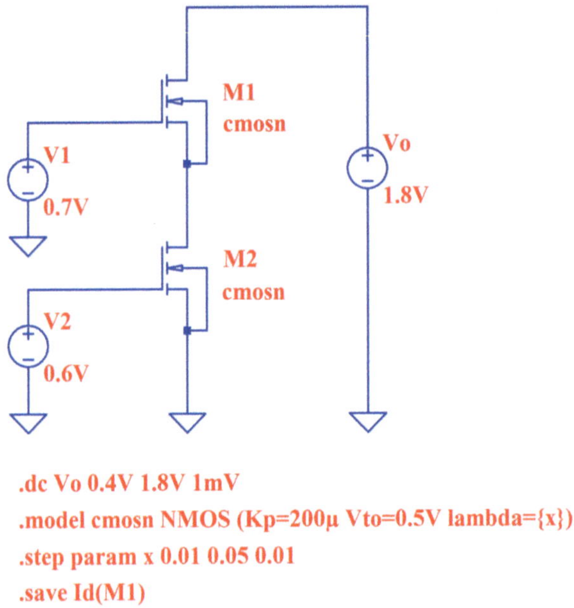

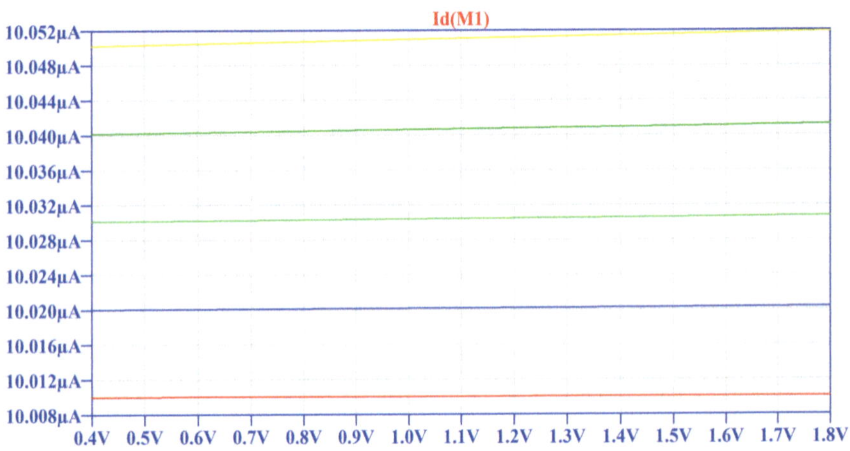

Fig. 1.53 Graphical representation of the characteristics $I_O(V_O)$, for multiple values of channel-length parameter λ_n

In order to evaluate the extremely important impact of setting the value of voltage source V_1 on the minimal output voltage V_{Omin}, expressed by Eq. (1.103), the simulation from Fig. 1.51 is made again for the new value of $V_1 = 1.8$ V (the new circuit is presented in Fig. 1.54). Setting a variation of the output voltage V_O between 1.2 and 1.8 V (a *DC sweep analysis* in LTspice), the graphical representation of the output current I_O is shown in Fig. 1.55 (the output characteristic of the MOS current source from Fig. 1.54). In order to directly view the result of the simulation, the

Fig. 1.54 MOS current source with *DC sweep analysis* for output voltage V_O

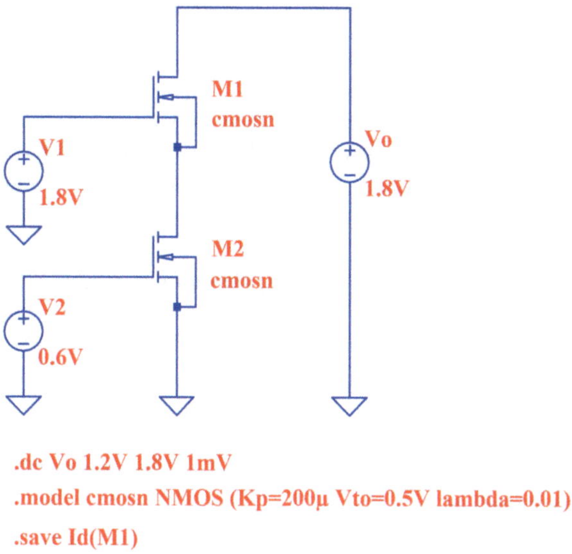

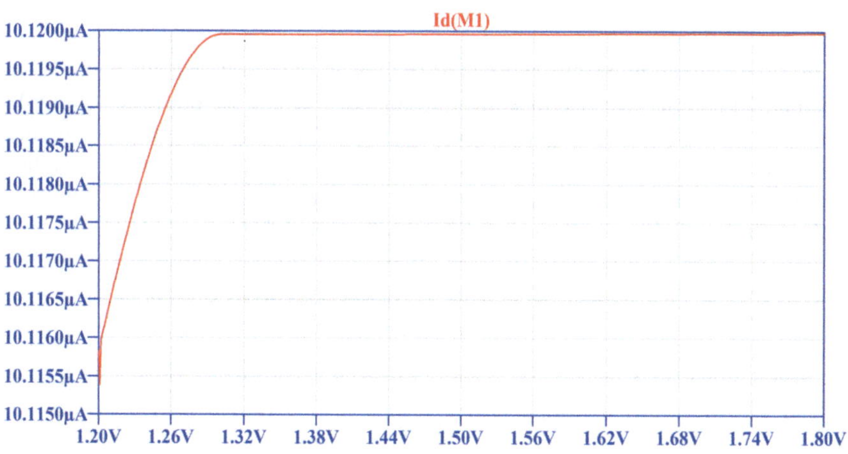

Fig. 1.55 Graphical representation of the output current I_O as a function of output voltage V_O

directive ".save Id(M1)" is introduced in the file, with Id(M1) representing the output current of the circuit I_O. The new theoretical value of V_{Omin} is now 1.8 V − 0.5 V = 1.3 V, confirmed by simulation. This new value, set for V_1 = 1.8 V, represents an intentionally wrong design example, which strongly increases the value of the voltage V_{Omin}, without obtaining any advantage in the circuit operation.

1.4.2 *Cascode MOS Current Mirror*

In order to strongly increase the output resistance R_O for the basic MOS current mirror presented in Fig. 1.16, the cascode version of the current mirror is presented in Fig. 1.56. The operation of this fundamental MOS current source is presented in detail in the classical textbooks of analog integrated circuits [1–4].

1.4.2.1 Output Current Expression I_O

The expression of the supply voltage can be written as follows:

$$V_{DD} = RI_{REF} + 2V_{GS1}. \tag{1.111}$$

Because of the biasing in saturation of transistor M1, the reference current can be expressed as

$$I_{REF} = \frac{K'_n}{2}\left(\frac{W}{L}\right)_1 (V_{GS1} - V_T)^2. \tag{1.112}$$

In order to determine the expression of I_{REF} current, the system of equations containing previous two relations must be solved. Because we have the $V_{GS1} \geq V_T$ criteria available for choosing the correct solution of the second-order equation that will be obtained by analysing these relations (because of the biasing in saturation region of transistor M1), the expression of the current I_{REF} from Eq. (1.112) must be replaced in Eq. (1.111), resulting in the following second-order equation:

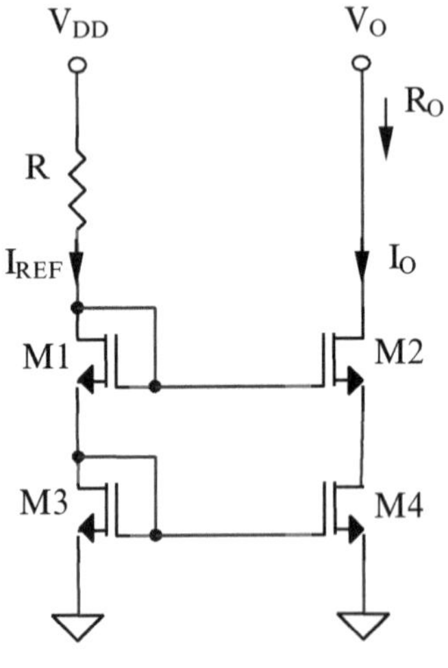

Fig. 1.56 Cascode MOS current mirror

$$V_{DD} = R\frac{K'_n}{2}\left(\frac{W}{L}\right)_1 (V_{GS1} - V_T)^2 + 2V_{GS1}. \tag{1.113}$$

The value of V_{GS1} is the solution of the previous second-order equation that fulfils the theoretical condition of biasing in saturation transistor M1, $V_{GS1} \geq V_T$. Replacing this value of V_{GS1} in Eq. (1.112), it results in the value of the reference current I_{REF}.

In order to determine the expression of the output current I_O and also the mirroring errors of the cascode MOS current mirror from Fig. 1.56, it is useful to compute the ratio I_O/I_{REF}:

$$\frac{I_O}{I_{REF}} = \frac{\frac{K'_n}{2}\left(\frac{W}{L}\right)_4 (V_{GS4} - V_T)^2 (1 + \lambda_n V_{DS4})}{\frac{K'_n}{2}\left(\frac{W}{L}\right)_3 (V_{GS3} - V_T)^2 (1 + \lambda_n V_{DS3})}. \tag{1.114}$$

The quantitative evaluation of the mirroring errors imposes the necessity of taking into account the channel-length modulation effect that affects the MOS transistor operation.

Because the circuit of the cascode MOS current mirror imposes $V_{GS3} = V_{GS4}$, $V_{DS3} = V_{GS3} = V_{GS}$ and $V_{DS4} = V_{GS1} + V_{GS3} - V_{GS2} = V_{GS}$ (all transistors are identical, and they are biased at approximately same drain currents, so their gate-source voltages are approximately equal), the previous ratio can be expressed as follows:

$$\frac{I_O}{I_{REF}} = \frac{\left(\frac{W}{L}\right)_4}{\left(\frac{W}{L}\right)_3} = 1. \tag{1.115}$$

This result has been obtained in the particular case of using identical transistors M3 and M4. For different aspect ratios of these transistors, there will exist a small difference between voltages V_{DS3} and V_{DS4}. In consequence, the cascode current mirror will be slightly affected in this case by the errors introduced by the channel-length modulation effect.

So, the mirroring factor for the cascode MOS current mirror is excellent, much better than the mirroring factor of the basic MOS current mirror, presented in Fig. 1.16. For the cascode MOS current mirror, this behavior is obtained because transistors M3 and M4 are operated at approximately same drain-source voltage (as a consequence of using transistors M1 and M2). The identity between drain-source voltages of transistors M3 and M4 practically cancels the errors introduced by the channel-length modulation effect. This very precise mirroring effect of the cascode MOS current mirror can be obtained for identical transistors, so for $I_O \cong I_{REF}$.

Fig. 1.57 Cascode MOS current mirror with *DC sweep analysis* for the output voltage V_O

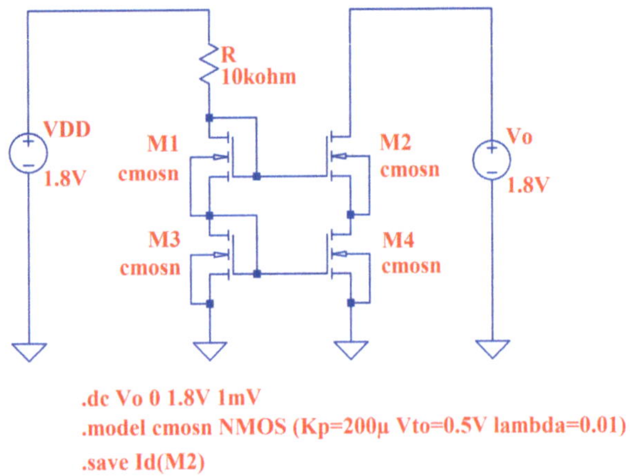

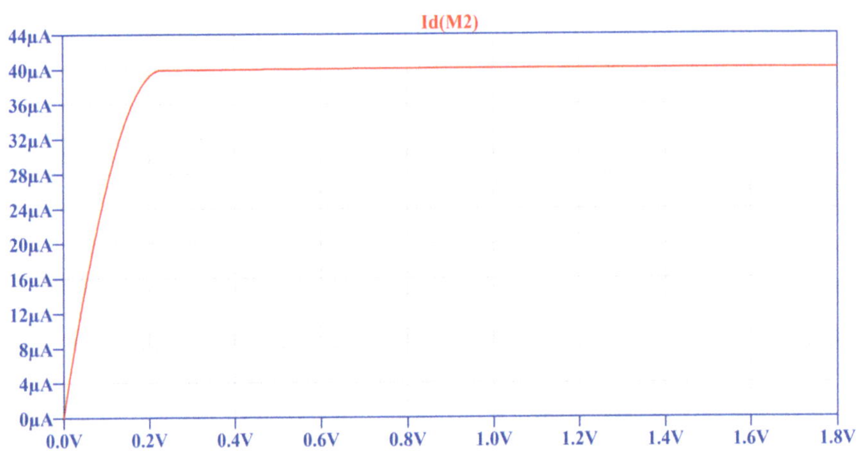

Fig. 1.58 Output characteristic $I_O(V_O)$ for the cascode current mirror presented in Fig. 1.57

Setting a variation of the output voltage V_O between 0 and 1.8 V (a *DC sweep analysis* in LTspice), the graphical representation of the output current I_O is shown in Fig. 1.58 (the output characteristic of the cascode MOS current mirror from Fig. 1.57). In order to directly view the result of the simulation, the directive ". save Id(M2)" is introduced in the file, with Id(M2) representing the output current of the circuit I_O.

The simulated value of the output current I_O is approximately 39.65 µA, very close to the theoretical value of 40 µA, expressed by Eq. (1.123).

The output current I_O is linearly increasing with the output voltage, but only for an output voltage superior to or equal to the minimal output voltage V_{Omin}, expressed by Eq. (1.126).

For a more facile evaluation with LTspice of the output resistance of the MOS current source from Fig. 1.57, the range of variation for the output voltage V_O is

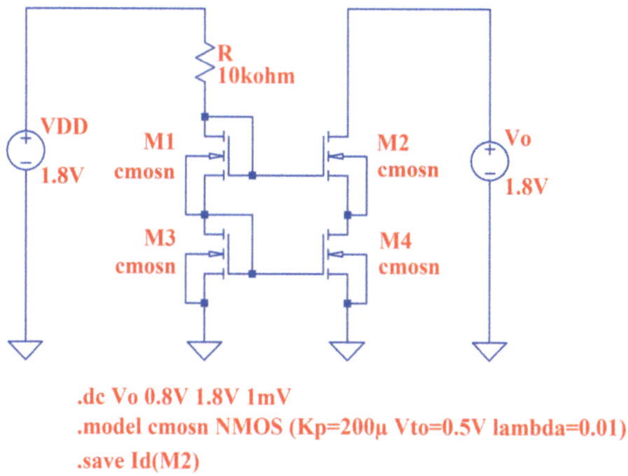

Fig. 1.59 Cascode MOS current mirror with limited *DC sweep analysis* for the output voltage V_O

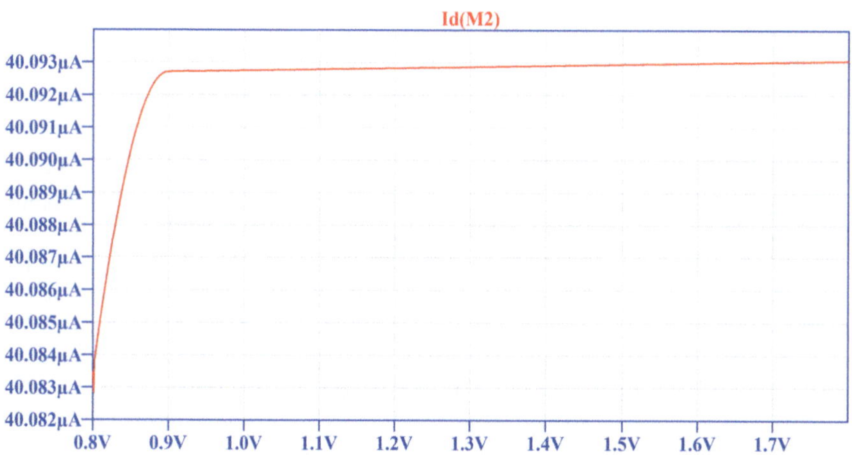

Fig. 1.60 Graphical representation of the characteristic $I_O(V_O)$, for a limited range of the output voltage V_O

limited (*DC sweep analysis* in LTspice, with the variation range for the output voltage V_O being set between 0.8 and 1.8 V). The circuit used for this simulation is presented in Fig. 1.59, with the graphical representation of the output current I_O being shown in Fig. 1.60. In order to directly view the result of the simulation, the directive ".save Id(M2)" is introduced in the file, with Id(M2) representing the output current of the circuit I_O.

In this simulation, the slope of the characteristic $I_O(V_O)$ is equal to $1/R_O$, with R_O being the output resistance of the MOS current source from Fig. 1.59. In consequence, the output resistance will have the following value:

Fig. 1.61 Circuit used for obtaining the graphical representation of the characteristics $I_O(V_O)$, for multiple values of the channel-length parameter λ_n

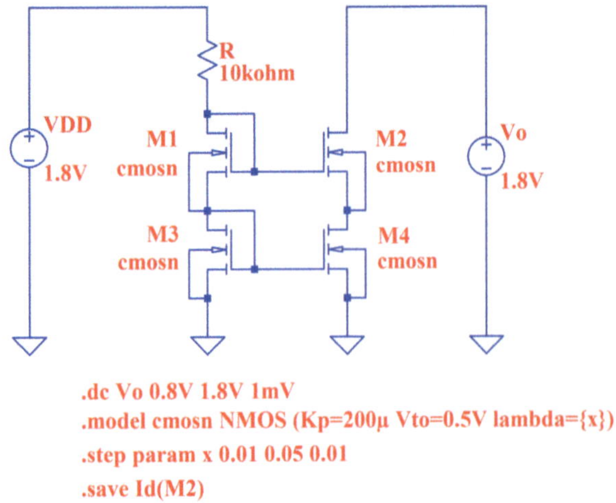

$$R_O = \frac{0.8\,\text{V}}{0.314\,\text{nA}} = 2.5\,\text{G}\Omega, \tag{1.127}$$

equal to the value of the theoretical output resistance, computed in Eq. (1.125).

The simulated value of the minimal output voltage is 0.9 V, equal to the theoretical value of V_{Omin}, expressed by Eq. (1.126).

In order to graphically represent the dependence of the output current I_O on the output voltage variation for multiple values of channel-length parameter λ_n (0.01, 0.02, 0.03, 0.04 and 0.05 V^{-1}), a *DC sweep analysis* is performed in LTspice, for a variation range of the output voltage V_O between 0.8 and 1.8 V and also a parametrical simulation for $x = \lambda_n$, between 0.01 and 0.05 V^{-1}, with a step of 0.01 V^{-1}. The circuit used for this simulation is presented in Fig. 1.61.

The parametrical simulation results for the output characteristics of the current source $I_O(V_O)$ are illustrated in Fig. 1.62. In order to directly view the results of the simulation, the directive ".save Id(M2)" is introduced in the file, with Id (M2) representing the output current of the circuit I_O.

In order to evaluate the dependence of the output current I_O on the supply voltage V_{DD}, a *DC sweep analysis* in LTspice is performed in Fig. 1.63 for a variation of the supply voltage V_{DD} between 0 and 1.8 V.

The simulation result for the characteristic $I_O(V_{DD})$ of the current source is illustrated in Fig. 1.64. In order to directly view the result of the simulation, the directive ".save Id(M2)" is introduced in the file, with Id(M2) representing the output current of the circuit I_O.

For a value of the supply voltage superior to a minimal value, the output current strongly increases when the supply voltage increases.

In Fig. 1.65, the dependence of the mirroring factor I_O/I_{REF} on the output voltage V_O is simulated. This factor is practically not affected by the channel-length modulation effect (see Eq. (1.115)).

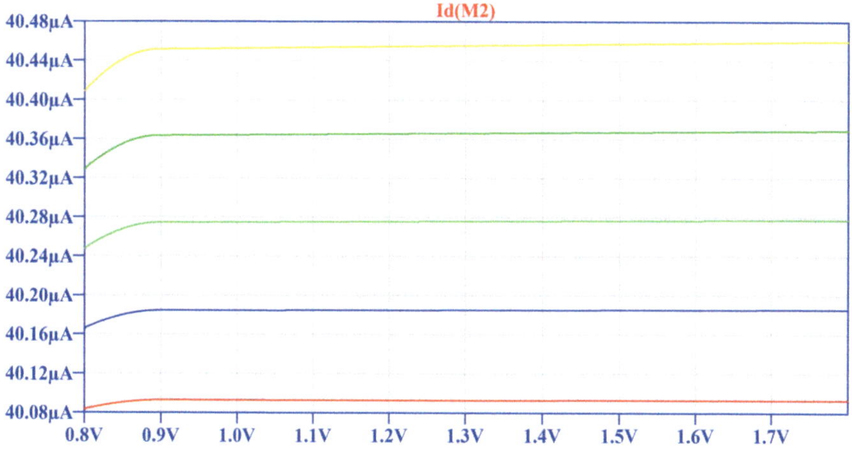

Fig. 1.62 Graphical representation of the characteristics $I_O(V_O)$, for multiple values of the channel-length parameter λ_n

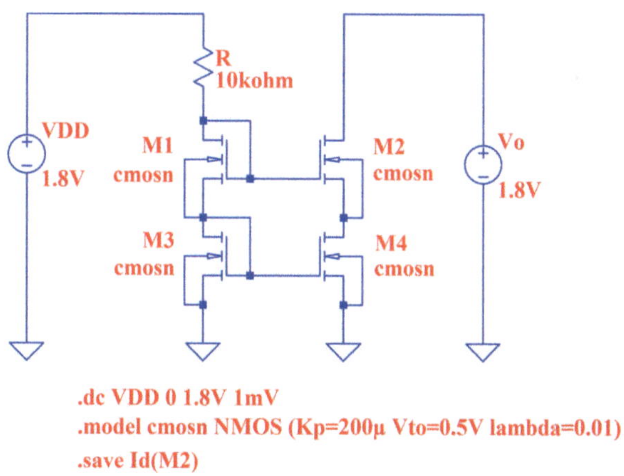

Fig. 1.63 Circuit used for obtaining the graphical representation of the dependence $I_O(V_{DD})$

1.4.3 Wide-Swing Cascode MOS Current Source (Version I)

In order to minimize the value of the minimal output voltage of a cascode current source V_{Omin}, the classical method is to bias the inferior output transistor at the limit of saturation. The wide-swing cascode MOS current source (version I) is presented in Fig. 1.66. The aspect ratios of all transistors are mentioned in the figure, with n being a parameter that will be set after finalizing the analysis of the current source. The inferior output transistor M5 must be biased at $V_{DS5} = V_{GS5} - V_T$ by using the loop containing the gate-source voltages of transistors M1 and M3.

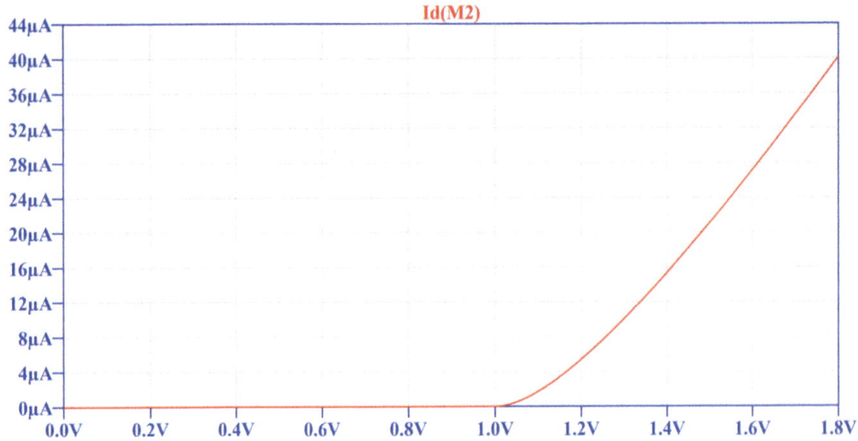

Fig. 1.64 Graphical representation of the characteristic $I_O(V_{DD})$

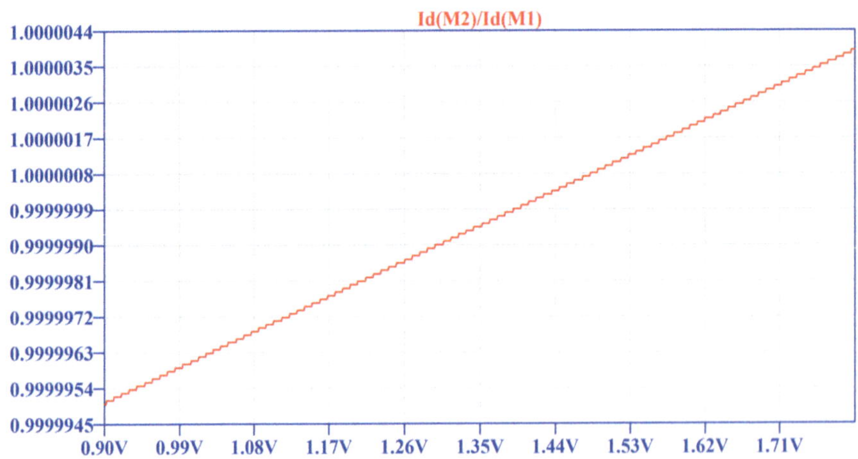

Fig. 1.65 The mirroring factor I_O/I_{REF} as a function of the output voltage V_O

1.4.3.1 Output Current Expression I_O

Transistors M4 and M5 form a basic current mirror. Because its transistors are identical, transistor M4 is biased at the current I and transistor M5 is biased at the current I_O, which results in $I_O \cong I$.

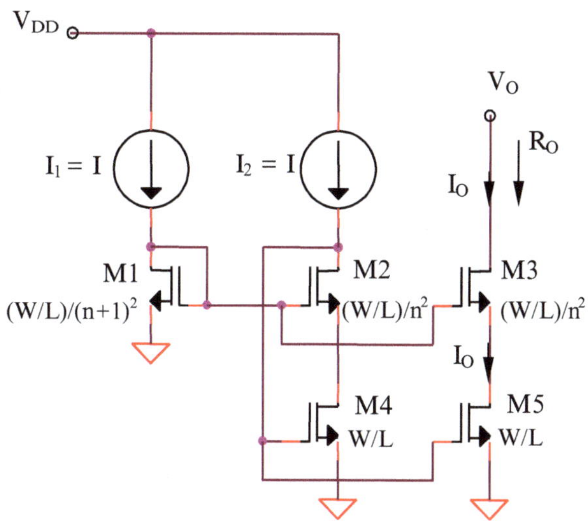

Fig. 1.66 Wide-swing cascode MOS current source (version I)

1.4.3.2 Output Resistance Expression R_O

Considering the expression of the dynamic resistance viewed in the drain terminal of the MOS transistor (determined using Eq. (A.10) from Annex), the output resistance of the current source from Fig. 1.66 can be expressed as follows:

$$R_O = r_{ds3}(1 + g_{m3}r_{ds5}) \cong g_{m3}r_{ds3}^2. \tag{1.128}$$

1.4.3.3 Minimal Output Voltage V_{Omin}

For biasing in the saturation region of all transistors from Fig. 1.66, it is possible to write that

$$I = \frac{K_n'}{2}\left(\frac{W}{L}\right)\frac{1}{(n+1)^2}(V_{GS1} - V_T)^2, \tag{1.129}$$

$$I = \frac{K_n'}{2}\left(\frac{W}{L}\right)\frac{1}{n^2}(V_{GS2(3)} - V_T)^2, \tag{1.130}$$

$$I = \frac{K_n'}{2}\left(\frac{W}{L}\right)(V_{GS4(5)} - V_T)^2. \tag{1.131}$$

From Eqs. (1.129) and (1.131), it results that

$$V_{GS1} - V_T = (n+1)\left(V_{GS4(5)} - V_T\right). \tag{1.132}$$

From Eqs. (1.130) and (1.131), it results that:

$$V_{GS2(3)} - V_T = n\left(V_{GS4(5)} - V_T\right). \tag{1.133}$$

The drain-source voltage of transistor M5 will have the following expression:

$$V_{DS5} = V_{GS1} - V_{GS3} = (V_{GS1} - V_T) - (V_{GS3} - V_T) = V_{GS5} - V_T. \tag{1.134}$$

So, transistor M5 will be biased at the limit of the saturation region, with this decision minimizing the minimal output voltage V_{Omin} for the cascode current source from Fig. 1.66:

$$V_{O\,min} = (V_{GS3} - V_T) + V_{DS5} = (n+1)(V_{GS5} - V_T) = (n+1)\sqrt{\frac{2I}{K'_n\left(\frac{W}{L}\right)}}. \tag{1.135}$$

The constant n must be set having as small as possible value, in order to minimize V_{Omin}.

1.4.3.4 Design Examples

In order to confirm the previously presented theoretical estimated results, the wide-swing cascode MOS current source (version I) from Fig. 1.66 is simulated with LTspice using the following values: $K'_n = 200\,\mu\text{A/V}^2$, $\lambda_n = 0.01\,\text{V}^{-1}$, $V_{Tn} = 0.5\,\text{V}$ and $V_{DD} = 1.8\,\text{V}$. MOS transistors have the aspect ratios given by $\left(\frac{W}{L}\right) = \frac{36\,\mu\text{m}}{0.5\,\mu\text{m}} = 72$ and $n = 2$, that is, $\left(\frac{W}{L}\right)_1 = \frac{4\,\mu\text{m}}{0.5\,\mu\text{m}} = 8$, $\left(\frac{W}{L}\right)_2 = \left(\frac{W}{L}\right)_3 = \frac{9\,\mu\text{m}}{0.5\,\mu\text{m}} = 18$ and $\left(\frac{W}{L}\right)_4 = \left(\frac{W}{L}\right)_5 = \frac{36\,\mu\text{m}}{0.5\,\mu\text{m}} = 72$. The current sources have the following values: $I_1 = I = 40\,\mu\text{A}$ and $I_2 = I = 40\,\mu\text{A}$.

The theoretical value of the output current for the wide-swing cascode MOS current source (version I) from Fig. 1.66 is $I_O \cong I = 40\,\mu\text{A}$.

In order to determine the theoretical value of the output resistance, the transconductance g_{m3} of transistor M3 must be determined:

$$g_{m3} = \sqrt{2 \times 200\,\mu\text{A/V}^2 \times 10 \times 40\,\mu\text{A}} = 400\,\mu\text{A/V}. \tag{1.136}$$

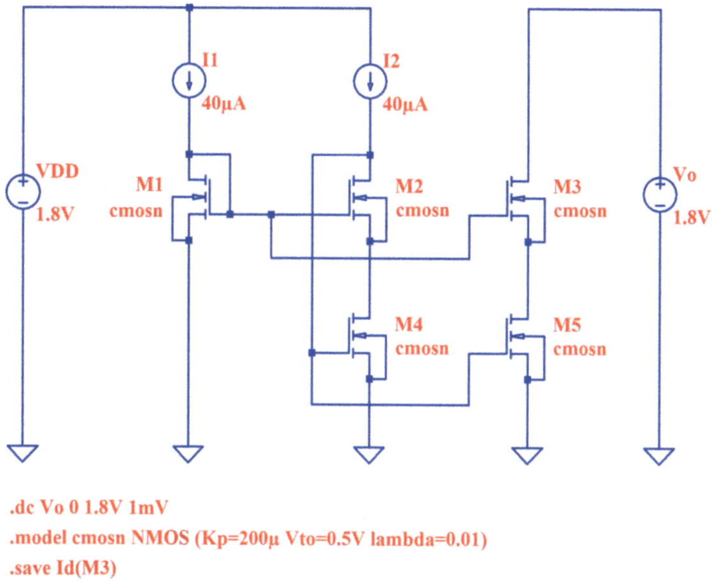

Fig. 1.67 Wide-swing cascode MOS current source (version I) with *DC sweep analysis* for output voltage V_O

The theoretical value of the output resistance can be obtained using Eq. (1.128):

$$R_O = 400\,\mu\text{A}/\text{V} \times \left(\frac{1}{0.01\,\text{V}^{-1} \times 40\,\mu\text{A}}\right)^2 = 2.5\,\text{G}\Omega. \tag{1.137}$$

The theoretical value of the minimal output voltage can be obtained using Eq. (1.135):

$$V_{O\,min} = (2+1)\sqrt{\frac{2 \times 40\,\mu\text{A}}{200\,\mu\text{A}/\text{V}^2 \times 72}} = 0.224\,\text{V}. \tag{1.138}$$

The circuit used for simulating the output characteristic $I_O(V_O)$ of the current source is shown in Fig. 1.67.

Setting a variation of the output voltage V_O between 0 and 1.8 V (a *DC sweep analysis* in LTspice), the graphical representation of the output current I_O is shown in Fig. 1.68 (the output characteristic of the wide-swing cascode MOS current source (version I) from Fig. 1.67). In order to directly view the result of the simulation, the directive ".save Id(M3)" is introduced in the file, with Id(M3) representing the output current of the circuit I_O.

The simulated value of the output current I_O is approximately 39.65 µA, very close to the theoretical value of 40 µA.

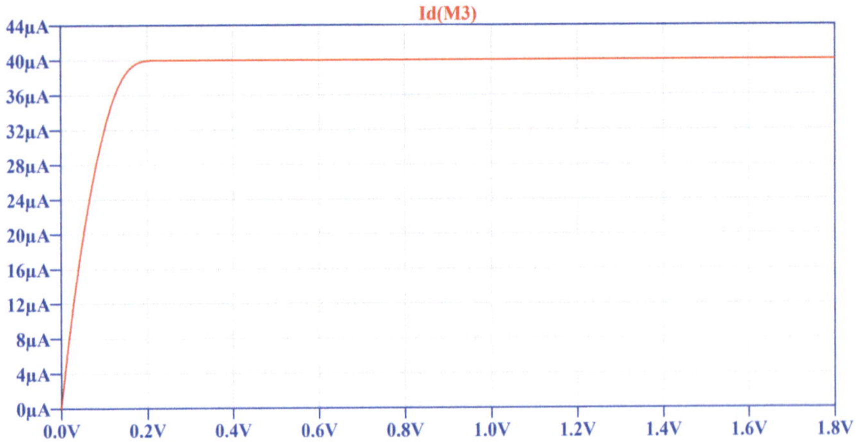

Fig. 1.68 Output characteristic $I_O(V_O)$ for the current source presented in Fig. 1.67

The output current I_O is linearly increasing with the output voltage, but only for an output voltage superior to or equal to the minimal output voltage V_{Omin}, expressed by Eq. (1.138).

For a more facile evaluation with LTspice of the output resistance of the MOS current source from Fig. 1.67, the range of variation for the output voltage V_O is limited (*DC sweep analysis* in LTspice, with the variation range for the output voltage V_O being set to be between 0.2 and 1.8 V). The circuit used for this simulation is presented in Fig. 1.69, with the graphical representation of the output current I_O being shown in Fig. 1.70. In order to directly view the result of the simulation, the directive ".save Id(M3)" is introduced in the file, with Id (M3) representing the output current of the circuit I_O.

In this simulation, the slope of the $I_O(V_O)$ characteristic is equal to $1/R_O$, with R_O being the output resistance of the MOS current source from Fig. 1.69. In consequence, the output resistance will have the following value:

$$R_O = \frac{1.8\,\text{V} - 1\,\text{V}}{0.3247\,\text{nA}} = \frac{0.8\,\text{V}}{0.3247\,\text{nA}} = 2.47\,\text{G}\Omega, \tag{1.139}$$

very close to the value of the theoretical output resistance, computed in Eq. (1.137).

The simulated value of the minimal output voltage is equal to 0.221 V, very close to the theoretical value of V_{Omin}, expressed by Eq. (1.138).

1.4.4 Wide-Swing Cascode MOS Current Source (Version II)

In order to minimize the value of the minimal output voltage of cascode current source V_{Omin}, a similar method is used, but having different circuit design. The

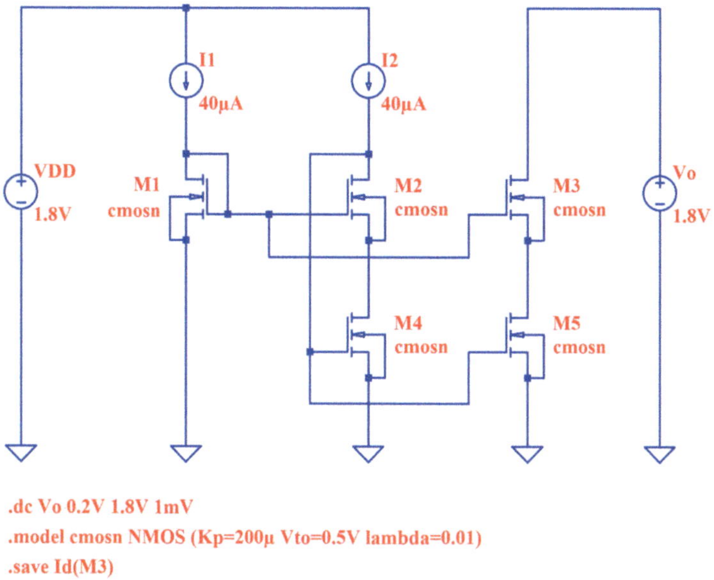

Fig. 1.69 Wide-swing cascode MOS current source (version I) with limited *DC sweep analysis* for the output voltage V_O

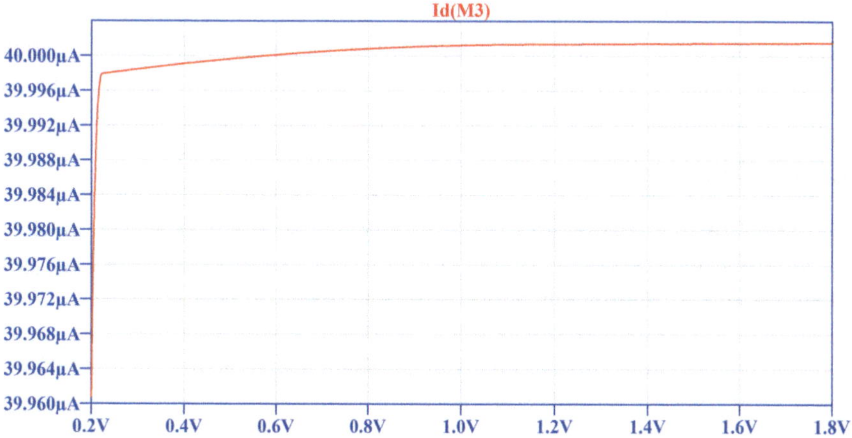

Fig. 1.70 Graphical representation of the characteristic $I_O(V_O)$, for a limited range of the output voltage V_O

wide-swing cascode MOS current source (version II) is presented in Fig. 1.71. The aspect ratios of all transistors are identical. The inferior output transistor M4 must be biased at $V_{DS4} = V_{GS4} - V_T$.

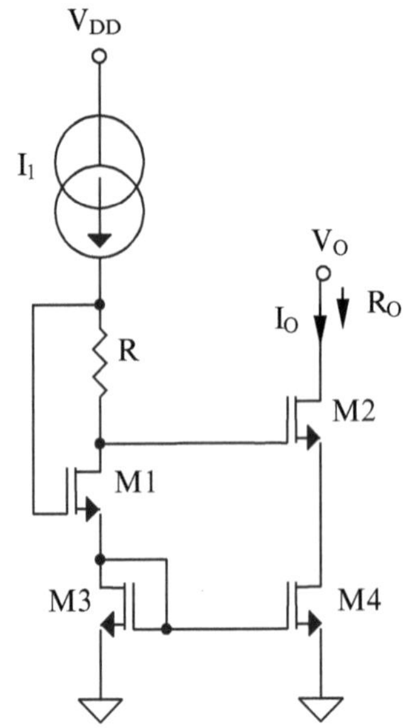

Fig. 1.71 Wide-swing cascode MOS current source (version II)

1.4.4.1 Output Current Expression I_O

Transistors M3 and M4 form a fundamental current mirror. Because its transistors are identical, transistor M3 is biased at the current I_1 and transistor M4 is biased at the current I_O, which results in $I_O \cong I_1$.

1.4.4.2 Output Resistance Expression R_O

Considering the expression of the dynamic resistance viewed in the drain terminal of the MOS transistor (determined using Eq. (A.10) from Annex), the output resistance of the current source from Fig. 1.71 can be expressed as follows:

$$R_O = r_{ds2}(1 + g_{m2}r_{ds4}) \cong g_{m2}r_{ds2}^2. \tag{1.140}$$

1.4.4.3 Minimal Output Voltage V_{Omin}

In order to bias transistor M4 at the limit of saturation, its drain-source voltage must be determined:

$$V_{DS4} = V_{GS3} + V_{GS1} - I_1R - V_{GS2}. \tag{1.141}$$

Because all transistors are identical and they are biased at the same drain current I_1, their gate-source voltages will be equal, so

$$V_{DS4} = V_{GS4} - I_1 R. \tag{1.142}$$

The biasing of transistor M4 at the limit of saturation can be obtained imposing the design condition that $I_1 R = V_T$. For $I_1 R > V_T$, it results that $V_{DS4} < V_{GS4} - V_T$, so transistor M4 will no longer be biased in saturation and, in consequence, the current source will no longer work. For $I_1 R < V_T$, it results that $V_{DS4} > V_{GS4} - V_T$, so the optimal value of V_{Omin} will no longer be achievable.

Imposing $I_1 R = V_T$, the expression of the minimal output voltage V_{Omin} for the wide-swing cascode MOS current source (version II) will be

$$V_{O\,min} = (V_{GS2} - V_T) + V_{DS4} = 2(V_{GS2} - V_T) = 2\sqrt{\frac{2I_1}{K'_n\left(\frac{W}{L}\right)}}. \tag{1.143}$$

1.4.4.4 Design Examples

In order to confirm the previously presented theoretical estimated results, the wide-swing cascode MOS current source (version II) from Fig. 1.71 is simulated with LTspice using the following values: $K'_n = 200\,\mu\text{A/V}^2$, $\lambda_n = 0.01\,\text{V}^{-1}$, $V_{Tn} = 0.5\,\text{V}$ and $V_{DD} = 1.8\,\text{V}$. MOS transistors have the aspect ratios given by $\left(\frac{W}{L}\right) = \frac{36\,\mu\text{m}}{0.5\,\mu\text{m}} = 72$. The values of the current source and of the resistor are $I_1 = 10\,\mu\text{A}$ and $R = 50\,\text{k}\Omega$ (so the condition $I_1 R = 10\,\mu\text{A} \times 50\,\text{k}\Omega = 0.5\,\text{V} = V_T$ is fulfilled).

The theoretical value of the output current for the wide-swing cascode MOS current source (version II) from Fig. 1.71 is $I_O \cong I_1 = 10\,\mu\text{A}$.

In order to determine the theoretical value of the output resistance, the transconductance g_{m2} of transistor M2 must be determined:

$$g_{m2} = \sqrt{2 \times 200\,\mu\text{A/V}^2 \times 10 \times 10\,\mu\text{A}} = 200\,\mu\text{A/V}. \tag{1.144}$$

The theoretical value of the output resistance can be obtained using Eq. (1.140):

$$R_O = 200\,\mu\text{A/V} \times \left(\frac{1}{0.01\,\text{V}^{-1} \times 10\,\mu\text{A}}\right)^2 = 20\,\text{G}\Omega. \tag{1.145}$$

The theoretical value of the minimal output voltage can be obtained using Eq. (1.143):

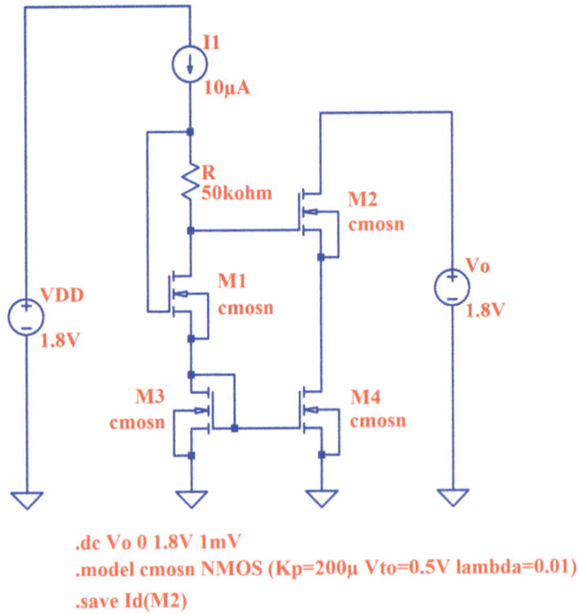

Fig. 1.72 Wide-swing cascode MOS current source (version II) with *DC sweep analysis* for the output voltage V_O

$$V_{O\,min} = 2\sqrt{\frac{2\times 10\,\mu\text{A}}{200\,\mu\text{A/V}^2\times 10}} = 0.2\,\text{V}. \tag{1.146}$$

The circuit used for simulating the output characteristic $I_O(V_O)$ of the wide-swing cascode MOS current source (version II) is shown in Fig. 1.72.

Setting a variation of the output voltage V_O between 0 and 1.8 V (a *DC sweep analysis* in LTspice), the graphical representation of the output current I_O is shown in Fig. 1.73 (the output characteristic of the wide-swing cascode MOS current source (version II) from Fig. 1.72). In order to directly view the result of the simulation, the directive ".save Id(M2)" is introduced in the file, with Id(M2) representing the output current of the circuit I_O.

The simulated value of the output current I_O is approximately 10 μA, equal to the theoretical value of 10 μA.

The output current I_O is linearly increasing with the output voltage, but only for an output voltage superior to or equal to the minimal output voltage V_{Omin}, expressed by Eq. (1.146).

For a more facile evaluation with LTspice of the output resistance of the MOS current source from Fig. 1.72, the range of variation for the output voltage V_O is limited (*DC sweep analysis* in LTspice, with the variation range for the output voltage V_O being set between 0.18 and 1.8 V). The circuit used for this simulation is presented in Fig. 1.74, with the graphical representation of the output current I_O being shown in Fig. 1.75. In order to directly view the result of the simulation, the

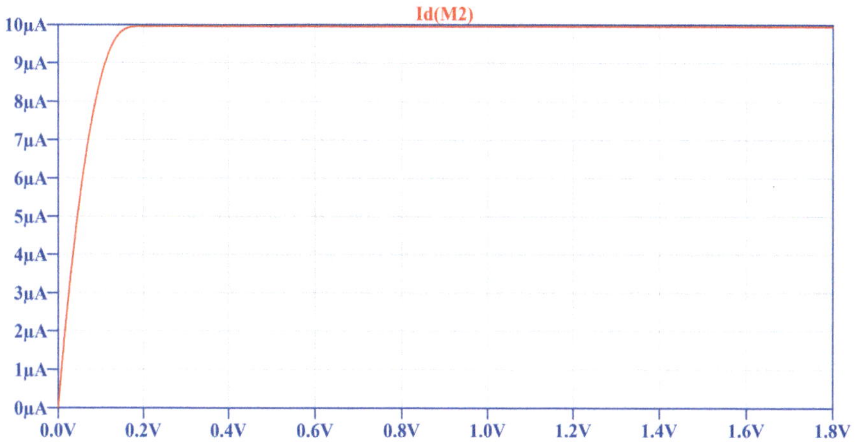

Fig. 1.73 Output characteristic $I_O(V_O)$ for the current source presented in Fig. 1.72

Fig. 1.74 Wide-swing cascode MOS current source (version II) with limited *DC sweep analysis* for the output voltage V_O

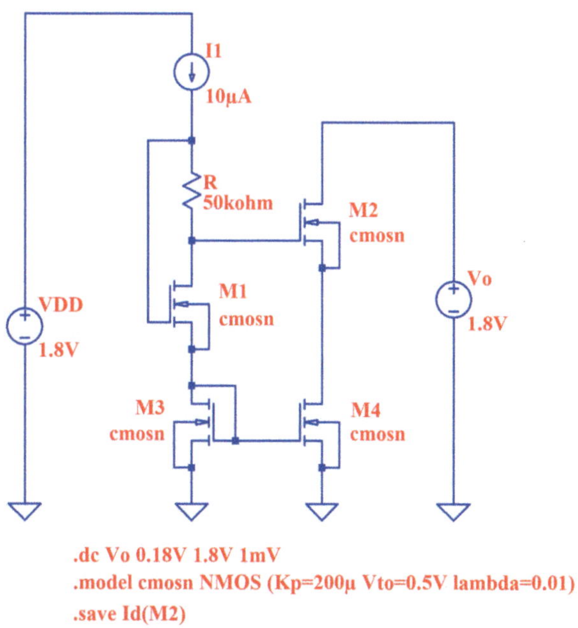

directive ".save Id(M2)" is introduced in the file, with Id(M2) representing the output current of the circuit I_O.

In this simulation, the slope of the $I_O(V_O)$ characteristic is equal to $1/R_O$, with R_O being the output resistance of the MOS current source from Fig. 1.74. In consequence, the output resistance will have the following value:

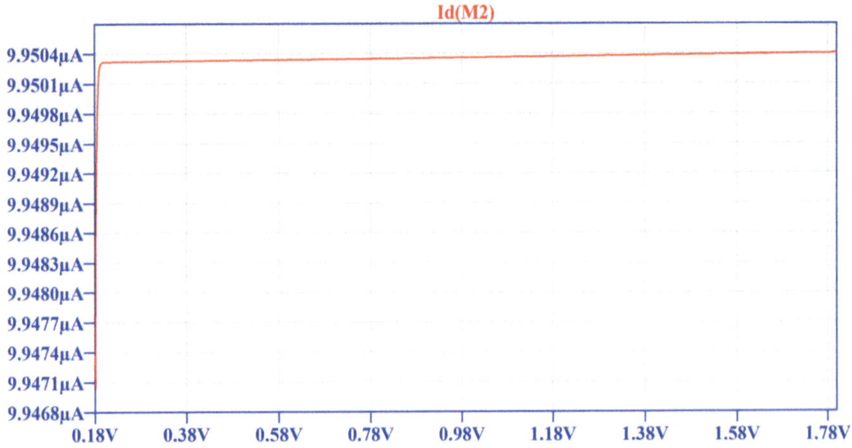

Fig. 1.75 Graphical representation of the characteristic $I_O(V_O)$, for a limited range of the output voltage V_O

$$R_O = \frac{1.8\,\text{V} - 0.8\,\text{V}}{48.7\,\text{pA}} = \frac{1\,\text{V}}{48.7\,\text{pA}} = 20.5\,\text{G}\Omega, \tag{1.147}$$

very close to the value of the theoretical output resistance, computed in Eq. (1.145).

The simulated value of the minimal output voltage is 0.2 V, equal to the theoretical value of V_{Omin}, expressed by Eq. (1.146).

1.4.5 *Wide-Swing Cascode MOS Current Source (Version III)*

The wide-swing cascode MOS current source (version III) is presented in Fig. 1.76. The aspect ratios of all transistors are identical and equal to $\left(\frac{W}{L}\right)$, except transistor M1, which must have an aspect ratio four times smaller than the aspect ratios of all other transistors, $\frac{1}{4}\left(\frac{W}{L}\right)$.

1.4.5.1 Output Current Expression I_O

Transistors M3, M4 and M6 form a basic current mirror. Because its transistors are identical, the output current I_O will be approximatively equal to the current I_{REF}. The reference current I_{REF} can be expressed as

Fig. 1.76 Wide-swing cascode MOS current source (version III)

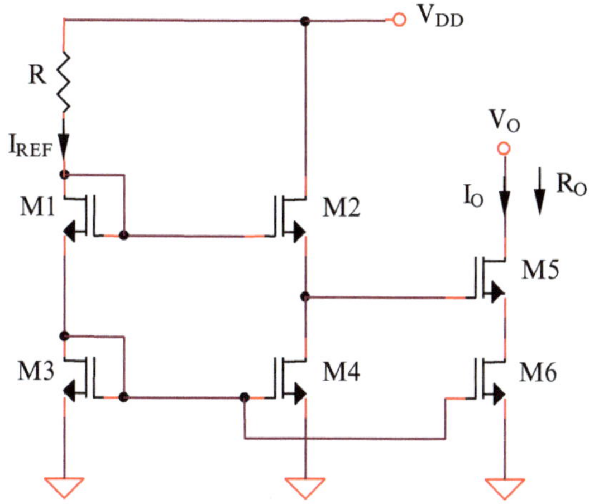

$$I_{REF} \cong \frac{K_n'}{2}\left(\frac{W}{L}\right)(V_{GS3} - V_T)^2. \tag{1.148}$$

The expression of the supply voltage can be written as follows:

$$V_{DD} = RI_{REF} + V_{GS1} + V_{GS3}. \tag{1.149}$$

Because transistors M1 and M3 are biased in saturation region, their gate-source voltages can be expressed as

$$V_{GS1} = V_T + \sqrt{\frac{2I_{REF}}{K_n' \frac{1}{4}\left(\frac{W}{L}\right)}} = V_T + 2\sqrt{\frac{2I_{REF}}{K_n'\left(\frac{W}{L}\right)}} \tag{1.150}$$

and

$$V_{GS3} = V_T + \sqrt{\frac{2I_{REF}}{K_n'\left(\frac{W}{L}\right)}}. \tag{1.151}$$

From Eqs. (1.150) and (1.151), the mathematical relation between these two gate-source voltages results:

$$V_{GS1} = 2V_{GS3} - V_T. \tag{1.152}$$

Replacing Eqs. (1.148) and (1.152) in Eq. (1.149), it could be concluded that

$$V_{DD} = R\frac{K'_n}{2}\left(\frac{W}{L}\right)(V_{GS3} - V_T)^2 + V_{GS3} + (2V_{GS3} - V_T), \tag{1.153}$$

equivalent to

$$V_{DD} + V_T = R\frac{K'_n}{2}\left(\frac{W}{L}\right)(V_{GS3} - V_T)^2 + 3V_{GS3}. \tag{1.154}$$

The value of the voltage V_{GS3} is the solution of the previous second-order equation that fulfils the theoretical condition of biasing in saturation transistor M3, $V_{GS3} \geq V_T$. Replacing this value of V_{GS3} in Eq. (1.148) results in the value of the reference current I_{REF} and also the value of the output current, $I_O \cong I_{REF}$.

1.4.5.2 Output Resistance Expression R_O

Considering the expression of the dynamic resistance viewed in the drain terminal of the MOS transistor (determined using Eq. (A.10) from Annex), the output resistance of the current source from Fig. 1.76 can be expressed as follows:

$$R_O = r_{ds5}(1 + g_{m5}r_{ds6}) \cong g_{m5}r_{ds5}^2. \tag{1.155}$$

1.4.5.3 Minimal Output Voltage V_{Omin}

In order to bias transistor M6 at the limit of saturation, its drain-source voltage must be determined:

$$V_{DS6} = V_{GS1} + V_{GS3} - V_{GS2} - V_{GS5}, \tag{1.156}$$

where the previous gate-source voltages can be expressed as follows:

$$V_{GS1} = V_T + \sqrt{\frac{2I_{REF}}{K'_n\frac{1}{4}\left(\frac{W}{L}\right)}} = V_T + 2\sqrt{\frac{2I_{REF}}{K'_n\left(\frac{W}{L}\right)}} \tag{1.157}$$

and

$$V_{GS3} = V_{GS2} = V_{GS5} = V_T + \sqrt{\frac{2I_{REF}}{K'_n\left(\frac{W}{L}\right)}}. \tag{1.158}$$

Replacing Eqs. (1.157) and (1.158) in Eq. (1.156) results in

$$V_{DS6} = \sqrt{\frac{2I_{REF}}{K_n'\left(\frac{W}{L}\right)}}. \tag{1.159}$$

Considering the squaring characteristic of transistor M6 biased in saturation region, it is possible to write that

$$V_{GS6} - V_T = \sqrt{\frac{2I_{REF}}{K_n'\left(\frac{W}{L}\right)}}. \tag{1.160}$$

In conclusions, $V_{DS6} = V_{GS6} - V_T$, so transistor M6 is biased at the limit of the saturation region, this minimizing the value of the voltage V_{Omin}. In consequence, the expression of the minimal output voltage V_{Omin} for the wide-swing cascode MOS current source (version III) will be

$$V_{O\,min} = (V_{GS5} - V_T) + V_{DS6} = 2(V_{GS5} - V_T) = 2\sqrt{\frac{2I_{REF}}{K_n'\left(\frac{W}{L}\right)}}. \tag{1.161}$$

1.4.5.4 Design Examples

In order to confirm the previously presented theoretical estimated results, the wide-swing cascode MOS current source (version III) from Fig. 1.76 is simulated with LTspice using the following values: $K_n' = 200\,\mu A/V^2$, $\lambda_n = 0.01\,V^{-1}$, $V_{Tn} = 0.5\,V$ and $V_{DD} = 1.8\,V$. MOS transistors have the aspect ratios given by $\left(\frac{W}{L}\right)_1 = \frac{5\,\mu m}{2\,\mu m} = 2.5$ and $\left(\frac{W}{L}\right)_2 = \left(\frac{W}{L}\right)_3 = \left(\frac{W}{L}\right)_4 = \left(\frac{W}{L}\right)_5 = \left(\frac{W}{L}\right)_6 = \frac{5\,\mu m}{0.5\,\mu m} = 10$ respectively, while the value of the resistor is $R = 10\,k\Omega$.

The theoretical value of the output current for the wide-swing cascode MOS current source (version III) from Fig. 1.76 could be determined using Eq. (1.154):

$$\frac{200\,\mu A/V^2}{2} \times 10 \times 10\,k\Omega \times (V_{GS3} - 0.5\,V)^2 + 3V_{GS3} = 2.3\,V, \tag{1.162}$$

resulting in the value $V_{GS3} = 0.67\,V$, corresponding to a biasing in saturation region of transistor M3. Replacing this value in Eq. (1.148) results in

$$I_O \cong \frac{200\,\mu A/V^2}{2} \times 10 \times (0.67\,V - 0.5\,V)^2 = 28.9\,\mu A. \tag{1.163}$$

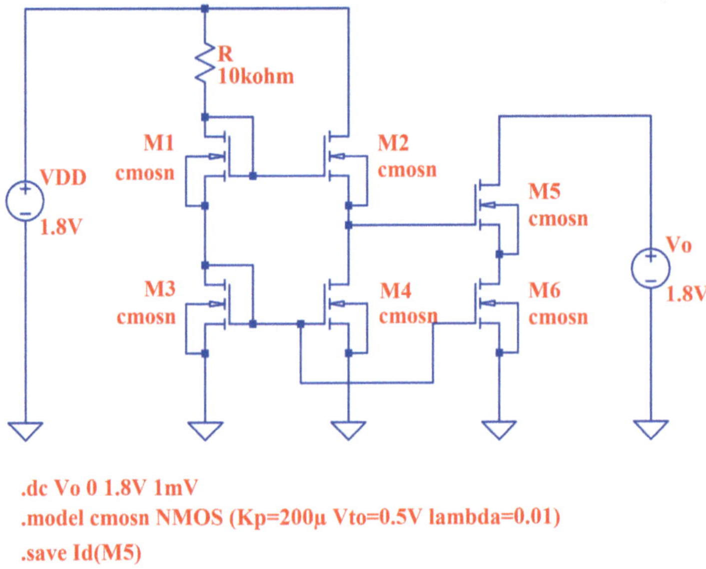

Fig. 1.77 Wide-swing cascode MOS current source (version III) with *DC sweep analysis* for output voltage V_O

In order to determine the theoretical value of the output resistance, the transconductance g_{m5} of transistor M5 must be determined:

$$g_{m5} = \sqrt{2 \times 200\,\mu\text{A/V}^2 \times 10 \times 28.9\,\mu\text{A}} = 340\,\mu\text{A/V}. \tag{1.164}$$

The theoretical value of the output resistance can be obtained using Eq. (1.155):

$$R_O = 340\,\mu\text{A/V} \times \left(\frac{1}{0.01\,\text{V}^{-1} \times 28.9\,\mu\text{A}}\right)^2 = 4.07\,\text{G}\Omega. \tag{1.165}$$

The theoretical value of the minimal output voltage can be obtained using Eq. (1.161):

$$V_{O\,min} = 2\sqrt{\frac{2 \times 28.9\,\mu\text{A}}{200\,\mu\text{A/V}^2 \times 10}} = 0.34\,\text{V}. \tag{1.166}$$

The circuit used for simulating the output characteristic $I_O(V_O)$ of the wide-swing MOS current source (version III) is shown in Fig. 1.77.

Setting a variation of the output voltage V_O between 0 and 1.8 V (a *DC sweep analysis* in LTspice), the graphical representation of the output current I_O is shown in Fig. 1.78 (the output characteristic of the wide-swing cascode MOS current source

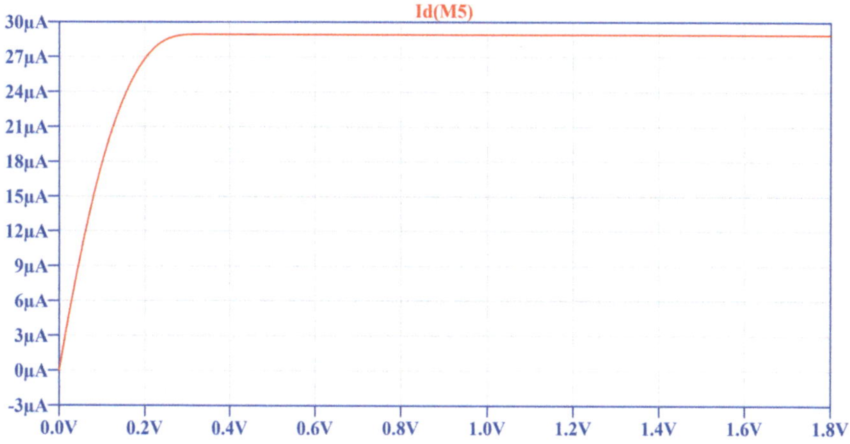

Fig. 1.78 Output characteristic $I_O(V_O)$ for the current source presented in Fig. 1.77

(version III) from Fig. 1.77). In order to directly view the result of the simulation, the directive ".save Id(M5)" is introduced in the file, with Id(M5) representing the output current of the circuit I_O.

The simulated value of the output current I_O is approximately 28.91 µA, very close to the theoretical value of 28.9 µA.

The output current I_O is linearly increasing with the output voltage, but only for an output voltage superior to or equal to the minimal output voltage V_{Omin}, expressed by Eq. (1.166).

For a more facile evaluation with LTspice of the output resistance of the MOS current source from Fig. 1.77, the range of variation for the output voltage V_O is limited (*DC sweep analysis* in LTspice, with the variation range for the output voltage V_O being set between 0.3 and 1.8 V). The circuit used for this simulation is presented in Fig. 1.79, with the graphical representation of the output current I_O being shown in Fig. 1.80. In order to directly view the result of the simulation, the directive ".save Id(M5)" is introduced in the file, with Id(M5) representing the output current of the circuit I_O.

In this simulation, the slope of the characteristic $I_O(V_O)$ is equal to $1/R_O$, with R_O being the output resistance of the MOS current source from Fig. 1.79. In consequence, the output resistance will have the following value:

$$R_O = \frac{1.8\,\text{V} - 0.8\,\text{V}}{242.7\,\text{pA}} = \frac{1\,\text{V}}{242.7\,\text{pA}} = 4.12\,\text{G}\Omega, \tag{1.167}$$

very close to the value of the theoretical output resistance, computed in Eq. (1.165).

The simulated value of the minimal output voltage is 0.34 V, equal to the theoretical value of V_{Omin}, expressed by Eq. (1.166).

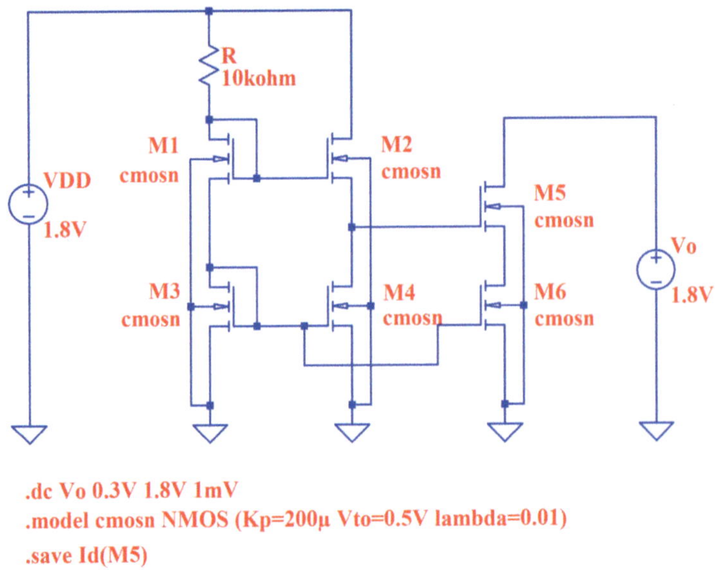

Fig. 1.79 Wide-swing cascode MOS current source (version III) with limited *DC sweep analysis* for the output voltage V_O

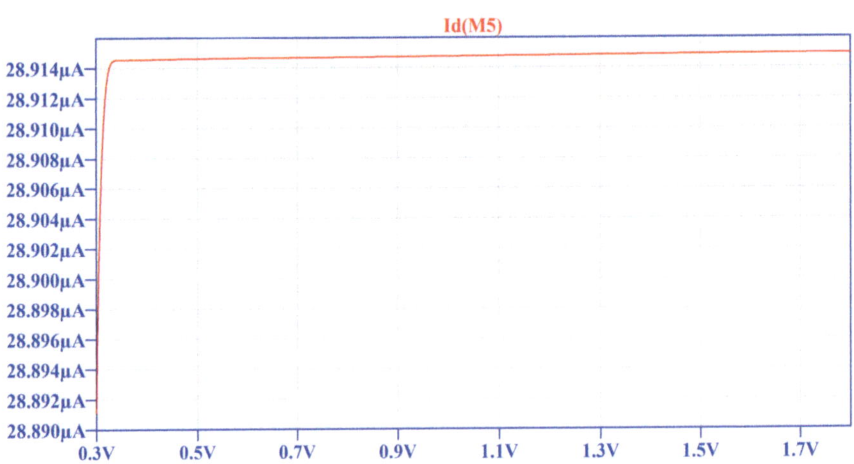

Fig. 1.80 Graphical representation of the characteristic $I_O(V_O)$, for a limited range of the output voltage V_O

1.4.6 Low-Voltage Wide-Swing Cascode MOS Current Source

A possible realization of a low-voltage wide-swing cascode MOS current source is presented in Fig. 1.81. The aspect ratios of all transistors are equal. The advantage of

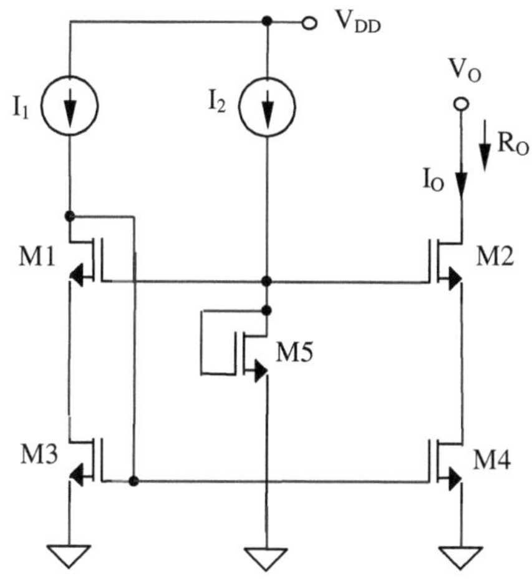

Fig. 1.81 Low-voltage wide-swing cascode MOS current source

this cascode current source in comparison to other similar wide-swing cascode current sources is mainly related to its much smaller minimal supply voltage V_{DDmin} (because of the connection between the drain of transistor M1 and the gate of transistor M3).

In order to bias in saturation region all transistors from Fig. 1.81, proper values of the current sources I_1 and I_2 must be set.

The condition of biasing in saturation region of transistor M1 can be written as $V_{DS1} \geq V_{GS1} - V_T$

$$V_{GS3} - (V_{GS5} - V_{GS1}) \geq V_{GS1} - V_T, \tag{1.168}$$

equivalent to

$$V_{GS3} \geq V_{GS5} - V_T. \tag{1.169}$$

Replacing the expressions of gate-source voltages as functions of drain currents results in

$$V_T + \sqrt{\frac{2I_1}{K'_n\left(\frac{W}{L}\right)}} \geq \sqrt{\frac{2I_2}{K'_n\left(\frac{W}{L}\right)}}. \tag{1.170}$$

The condition of biasing in saturation region of transistor M3 can be written as $V_{DS3} \geq V_{GS3} - V_T$

$$V_{GS5} - V_{GS1} \geq V_{GS3} - V_T. \tag{1.171}$$

Replacing the expressions of gate-source voltages as functions of the drain currents results in

$$\sqrt{\frac{2I_2}{K_n'\left(\frac{W}{L}\right)}}-\sqrt{\frac{2I_1}{K_n'\left(\frac{W}{L}\right)}}\geq\sqrt{\frac{2I_1}{K_n'\left(\frac{W}{L}\right)}}, \tag{1.172}$$

equivalent to

$$I_2\geq 4I_1. \tag{1.173}$$

1.4.6.1 Output Current Expression I_O

Transistors M3 and M4 form a basic current mirror. Because its transistors are identical, the output current I_O will be equal to the drain current of transistor M1, so $I_O \cong I_1$.

1.4.6.2 Output Resistance Expression R_O

Considering the expression of the dynamic resistance viewed in the drain terminal of the MOS transistor (determined using Eq. (A.10) from Annex), the output resistance of the current source from Fig. 1.81 can be expressed as follows:

$$R_O=r_{ds2}(1+g_{m2}r_{ds4})\cong g_{m2}r_{ds2}^2. \tag{1.174}$$

1.4.6.3 Minimal Output Voltage V_{Omin}

The expression of the minimal output voltage V_{Omin} for the low-voltage wide-swing cascode MOS current source is given by the condition of biasing in saturation region of transistor M2:

$$V_{O\,min}=V_{GS5}-V_{GS2}+(V_{GS2}-V_T)=V_{GS5}-V_T=\sqrt{\frac{2I_2}{K_n'\left(\frac{W}{L}\right)}}. \tag{1.175}$$

1.4.6.4 Design Examples

In order to confirm the previously presented theoretical estimated results, the low-voltage wide-swing cascode MOS current source from Fig. 1.81 is simulated with LTspice using the following values: $K_n'=200\,\mu A/V^2$, $\lambda_n = 0.01\,V^{-1}$, $V_{Tn} = 0.5\,V$ and $V_{DD} = 1.8\,V$. All transistors have the aspect ratios equal to

$\left(\frac{W}{L}\right) = \frac{5\,\mu\text{m}}{0.5\,\mu\text{m}} = 10$. The values of the current sources are $I_1 = 10\,\mu\text{A}$ and $I_2 = 80\,\mu\text{A}$ (values that fulfil conditions expressed by Eqs. (1.170) and (1.173)).

The theoretical value of the output current for the low-voltage wide-swing cascode MOS current source from Fig. 1.81 is $I_O \cong I_1 = 10\,\mu\text{A}$.

In order to determine the theoretical value of the output resistance, the transconductance g_{m2} of transistor M2 must be determined:

$$g_{m2} = \sqrt{2 \times 200\,\mu\text{A/V}^2 \times 10 \times 10\,\mu\text{A}} = 200\,\mu\text{A/V}. \tag{1.176}$$

The theoretical value of the output resistance can be obtained using Eq. (1.174):

$$R_O = 200\,\mu\text{A/V} \times \left(\frac{1}{0.01\,\text{V}^{-1} \times 10\,\mu\text{A}}\right)^2 = 20\,\text{G}\Omega. \tag{1.177}$$

The theoretical value of the minimal output voltage can be obtained using Eq. (1.175):

$$V_{O\,min} = \sqrt{\frac{2 \times 80\,\mu\text{A}}{200\,\mu\text{A/V}^2 \times 10}} = 0.283\,\text{V}. \tag{1.178}$$

The circuit used for simulating the output characteristic $I_O(V_O)$ of the current source is shown in Fig. 1.82.

Setting a variation of the output voltage V_O between 0 and 1.8 V (a *DC sweep analysis* in LTspice), the graphical representation of the output current I_O is shown in Fig. 1.83 (the output characteristic of the low-voltage wide-swing cascode MOS current source from Fig. 1.82). In order to directly view the result of the simulation, the directive ".save Id(M2)" is introduced in the file, with Id(M2) representing the output current of the circuit I_O.

The simulated value of the output current I_O is approximately 10 μA, equal to the theoretical value.

The output current I_O is linearly increasing with the output voltage, but only for an output voltage superior to or equal to the minimal output voltage V_{Omin}, expressed by Eq. (1.175).

For a more facile evaluation with LTspice of the output resistance of the MOS current source from Fig. 1.82, the range of variation for the output voltage V_O is limited (*DC sweep analysis* in LTspice, with the variation range for the output voltage V_O being set between 0.25 and 1.8 V). The circuit used for this simulation is presented in Fig. 1.84, with the graphical representation of the output current I_O being shown in Fig. 1.85. In order to directly view the result of the simulation, the directive ".save Id(M2)" is introduced in the file, with Id(M2) representing the output current of the circuit I_O.

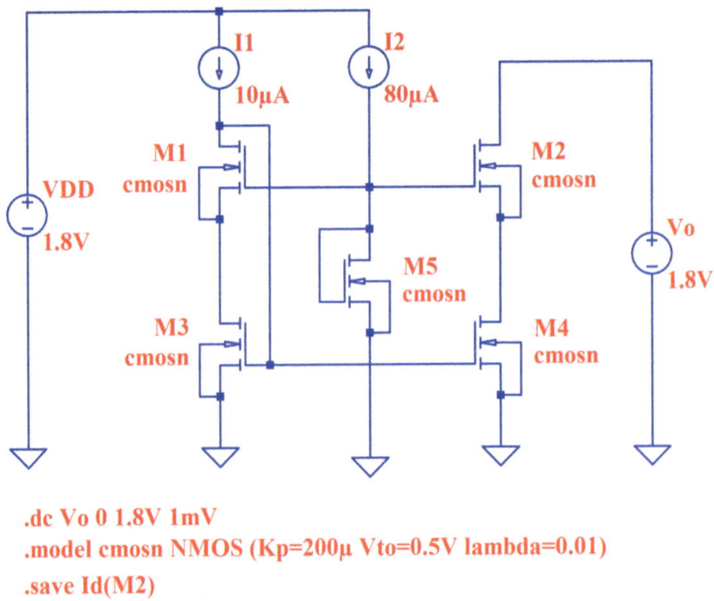

Fig. 1.82 Low-voltage wide-swing cascode MOS current source with *DC sweep analysis* for the output voltage V_O

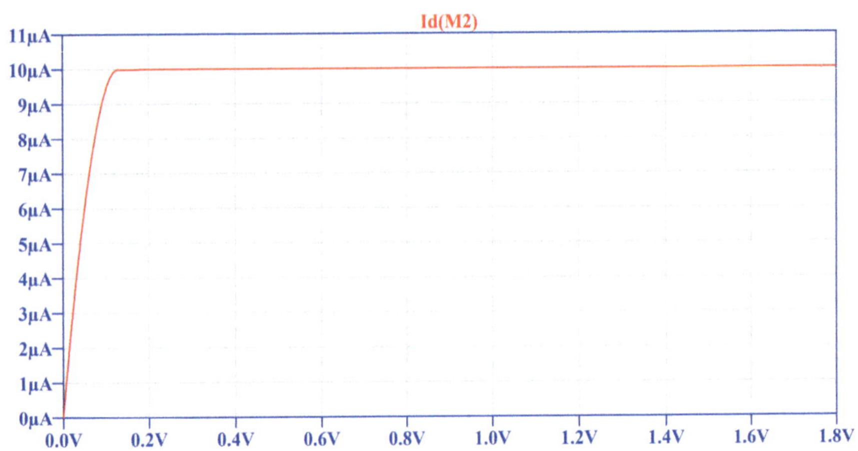

Fig. 1.83 Output characteristic $I_O(V_O)$ for the current source presented in Fig. 1.82

In this simulation, the slope of the characteristic $I_O(V_O)$ is equal to $1/R_O$, with R_O being the output resistance of the MOS current source from Fig. 1.84. In consequence, the output resistance will have the following value:

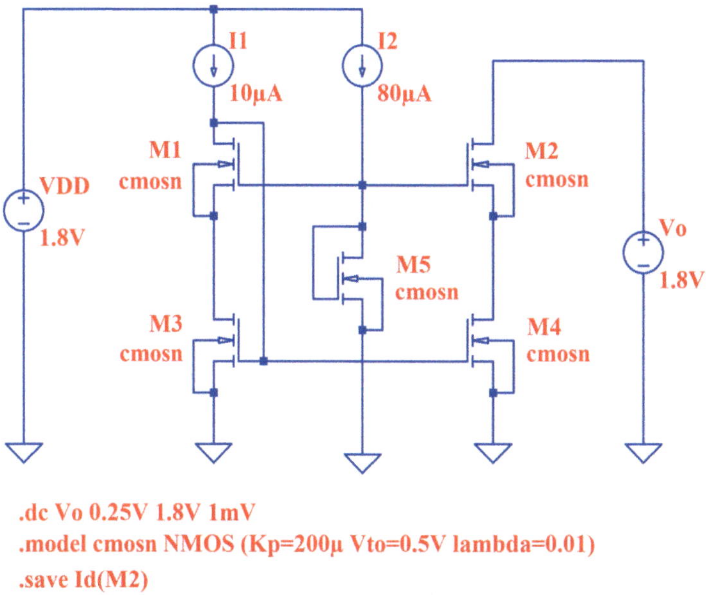

Fig. 1.84 Low-voltage wide-swing cascode MOS current source with limited *DC sweep analysis* for output voltage V_O

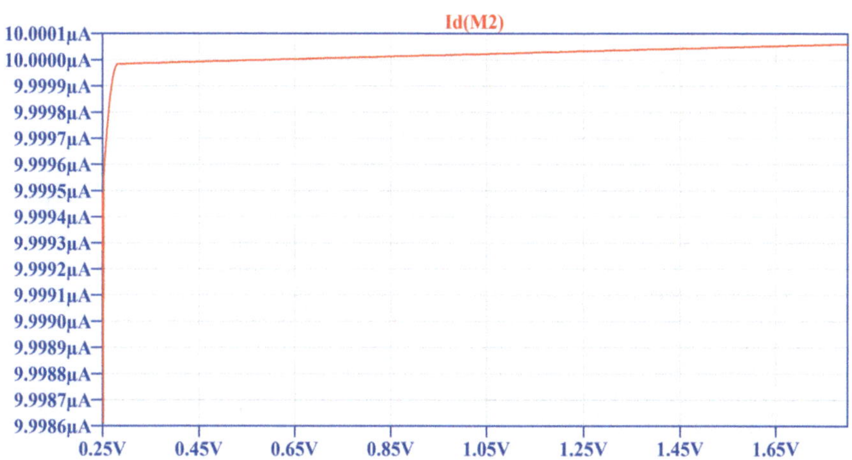

Fig. 1.85 Graphical representation of the characteristic $I_O(V_O)$, for a limited range of the output voltage V_O

$$R_O = \frac{1.8\,\text{V} - 0.3\,\text{V}}{75.93\,\text{pA}} = \frac{1.5\,\text{V}}{75.93\,\text{pA}} = 19.76\,\text{G}\Omega, \tag{1.179}$$

very close to the value of the theoretical output resistance, computed in Eq. (1.177).

The simulated value of the minimal output voltage is equal to 0.28 V, very close to the theoretical value of V_{Omin}, expressed by Eq. (1.178).

1.5 Comparison Between Basic Current Sources and Cascode Current Sources

A comparison between the previous basic current sources and cascode current sources is presented in Table 1.1. The results are obtained for a supply voltage $V_{DD} = 1.8$ V.

The most important conclusions of this comparison are the following:

- All cascode current sources have an output resistance R_O much greater than the output resistance of the basic current sources. The difference between these parameters is at least three orders of magnitude, being increased from some megaohms to some gigaohms or even tens of gigaohms.
- The cascode MOS current mirror has a relatively large value of the minimal output voltage V_{Omin} in comparison to all basic current sources. The value of V_{Omin} can be decreased to reasonable values, comparable to those of the basic current sources, by using specific wide-swing techniques (wide-swing cascode MOS current sources (versions I, II and III) and also low-voltage wide-swing cascode MOS current source).

Table 1.1 Comparison between the previous basic current sources and cascode current sources

Parameter/current source	Fig. no.	Output resistance R_O	Minimal output voltage V_{Omin} (V)
Basic current mirror	1.16	1.014 MΩ	0.314
MOS current source with increased output resistance (version I)	1.29	2.384 MΩ	0.314
MOS current source with increased output resistance (version II)	1.38	2.99 MΩ	0.393
Cascode MOS current mirror	1.56	2.5 GΩ	0.9
Wide-swing cascode MOS current source (version I)	1.66	2.5 GΩ	0.224
Wide-swing cascode MOS current source (version II)	1.71	20 GΩ	0.2
Wide-swing cascode MOS current source (version III)	1.76	4.07 GΩ	0.34
Low-voltage wide-swing cascode MOS current source	1.81	20 GΩ	0.283

1.6 Self-Biased Current Sources

Most of the previously presented current sources have a very important problem: the strong dependence of the output current on the supply voltage variations. In order to reduce this undesired dependence, the self-biasing principle could be used, with the result being a reduction with about two orders of magnitude of the dependence $I_O(V_{DD})$, in comparison to classical current sources.

1.6.1 Self-Biased CTAT Current Source

A complementary to absolute temperature (CTAT) current source represents a circuit that is able to generate a current that is linearly decreasing with temperature. A possible realization of such a circuit is presented in Fig. 1.86.

Transistors M3, M4 and M5 form a basic current mirror with identical transistors, so all currents from the circuit are approximately equal to I_O. Because $V_{GS1} = V_{GS2}$, it results that $V_{EB1} = I_O R_1$. So, the expression of the output current will be

$$I_O = \frac{V_{EB1}}{R_1}. \tag{1.180}$$

Because the emitter-base voltage is approximately decreasing with temperature with an approximate slope of $-2\,\text{mV/K}$ and neglecting the temperature dependence of resistor R_1, the output current I_O will have a CTAT behavior. The temperature coefficient of the output current can be expressed as follows:

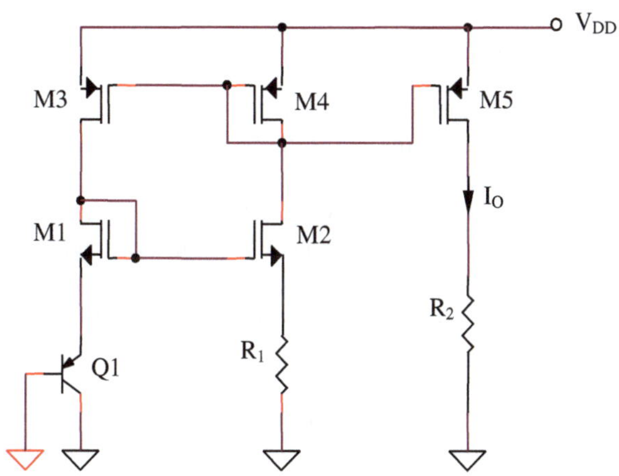

Fig. 1.86 Self-biased CTAT current source

$$TC_{I_O} = \frac{dI_O}{dT} = \frac{1}{R_1}\frac{dV_{EB1}}{dT}. \tag{1.181}$$

1.6.1.1 Design Examples

In order to confirm the previously presented theoretical estimated results, the self-biased CTAT current source from Fig. 1.86 is simulated with LTspice using the following values: $K'_n = 200\,\mu A/V^2$, $K'_p = 100\,\mu A/V^2$, $\lambda_n = 0.01\,V^{-1}$, $\lambda_p = 0.02\,V^{-1}$, $V_{Tn} = 0.5\,V$, $V_{Tp} = -0.5\,V$ and $V_{DD} = 1.8\,V$. All transistors have the aspect ratios equal to $\left(\frac{W}{L}\right) = \frac{5\,\mu m}{0.5\,\mu m} = 10$, while resistors have the following values: $R_1 = 10\,k\Omega$ and $R_2 = 10\,k\Omega$.

The theoretical temperature coefficient of the output current can be obtained using Eq. (1.181):

$$TC_{I_O} = \frac{1}{10\,k\Omega}(-2\,mV/K) = -0.2\,\mu A/K. \tag{1.182}$$

For evaluating the temperature behavior of the output current I_O with LTspice, a *DC sweep analysis* is performed, with the variation range for the temperature being set between 0 and 100 °C. The circuit used for simulating the temperature dependence of the output current for the self-biased CTAT current source is presented in Fig. 1.87, with the graphical representation $I_O(T)$ being shown in Fig. 1.88. In order to directly view the result of the simulation, the directive ".save I(R2)" is introduced in the file, with I(R2) representing the output current of the circuit I_O.

The simulated value of the temperature coefficient of the output current is

$$TC_{I_O} = \frac{69.41\,\mu A - 50.56\,\mu A}{100\,K} = -0.19\,\mu A/K, \tag{1.183}$$

very close to the theoretical value obtained using Eq. (1.182).

For evaluating the dependence of the output current I_O on the supply voltage variations with LTspice, a *DC sweep analysis* is performed, with the variation range for the supply voltage being set between 0 and 3 V. The circuit used for this simulation is presented in Fig. 1.89, with the graphical representation of the dependence of the output current on the supply voltage variations being shown in Fig. 1.90. In order to directly view the result of the simulation, the directive ".save I(R2)" is introduced in the file, with I(R2) representing the output current of the circuit I_O.

The cause for obtaining a small, but non-zero dependence $I_O(V_{DD})$ (for a value of the supply voltage superior to a minimal value) is represented by the errors introduced by the current mirror M3–M4, as a consequence of the channel-length modulation effect.

Fig. 1.87 Circuit used for simulating the dependence $I_O(T)$

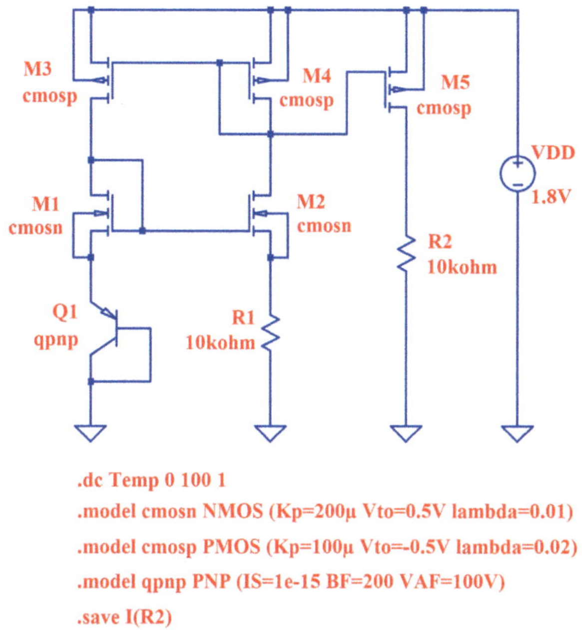

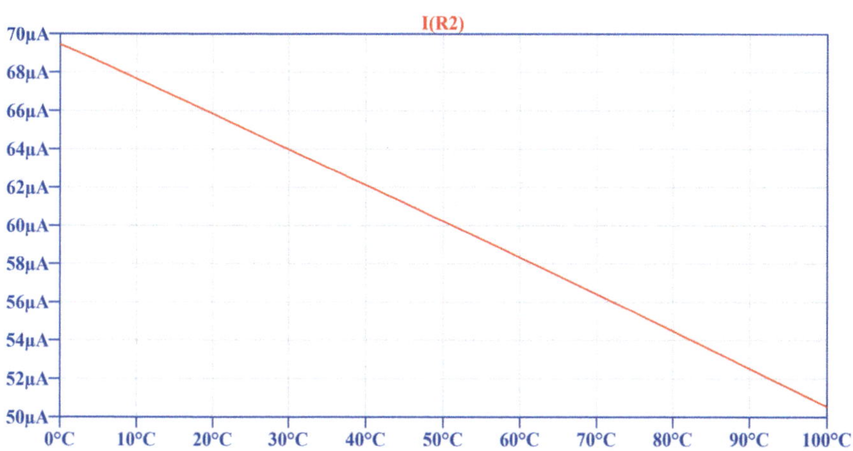

Fig. 1.88 The graphical representation of the dependence $I_O(T)$

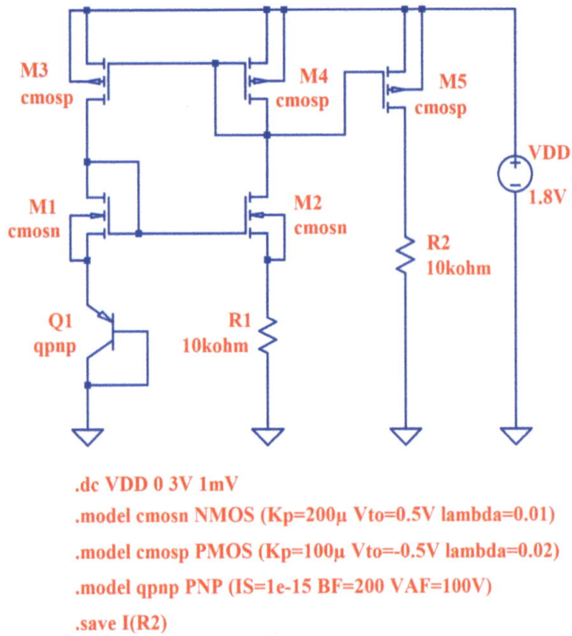

Fig. 1.89 Circuit used for simulating the dependence $I_O(V_{DD})$

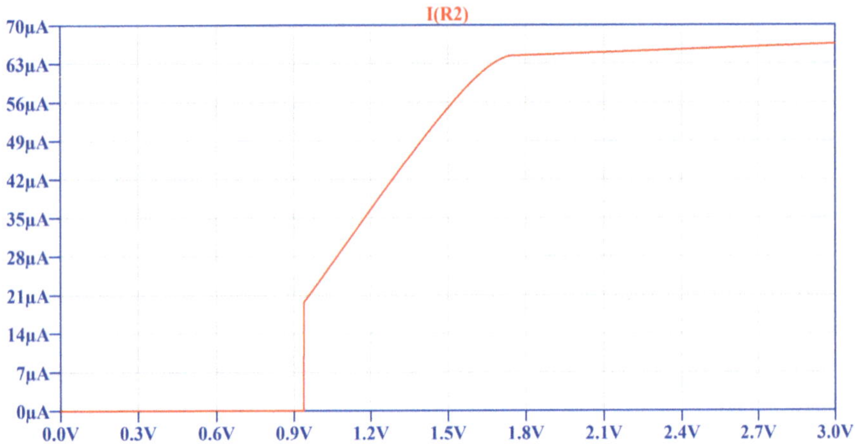

Fig. 1.90 Graphical representation of the dependence $I_O(V_{DD})$

1.6.2 Self-Biased PTAT Current Source

A proportional to absolute temperature (PTAT) current source represents a circuit that is able to generate a current that is linearly increasing with temperature. A possible realization of such a circuit is presented in Fig. 1.91.

Fig. 1.91 Self-biased PTAT current source

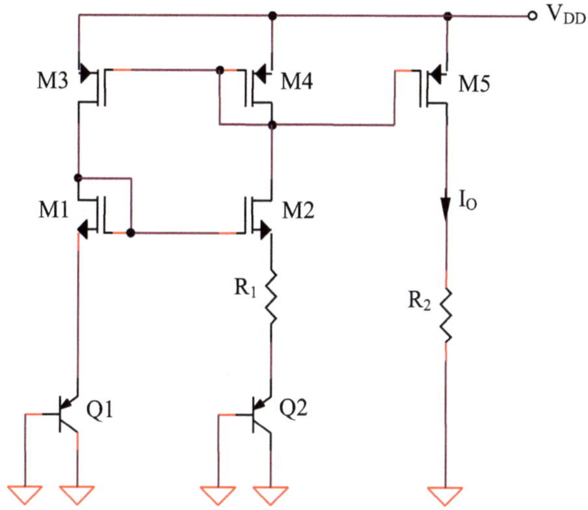

Transistors M3, M4 and M5 form a fundamental current mirror with identical transistors, so all currents from the circuit are approximately equal to I_O. Using identical transistors M1 and M2, their gate-source voltages will be equal, $V_{GS1} = V_{GS2}$, resulting that

$$V_{EB1} = I_O R_1 + V_{EB2}. \tag{1.184}$$

So, the expression of the output current will be

$$I_O = \frac{V_{EB1} - V_{EB2}}{R_1}. \tag{1.185}$$

Because the emitter-base voltage is logarithmically dependent on the collector current,

$$V_{EB} = V_{th} \ln\left(\frac{I_C}{I_S}\right), \tag{1.186}$$

where $V_{th} = \frac{kT}{q}$ represents the thermal voltage and I_S is the saturation current, the previous relation becomes

$$I_O = \frac{V_{th} \ln\left(\frac{I_{C1}}{I_{S1}}\right) - V_{th} \ln\left(\frac{I_{C2}}{I_{S2}}\right)}{R_1} = \frac{V_{th}}{R_1} \ln\left(\frac{I_{C1}}{I_{C2}} \frac{I_{S2}}{I_{S1}}\right). \tag{1.187}$$

The thermal voltage V_{th} is the same for both transistors Q1 and Q2, because they operate at the same temperature (the circuit is designed as integrated circuit).

Neglecting the base currents of transistors Q1 and Q2 and replacing the ratio of the saturation currents $\frac{I_{S2}}{I_{S1}}$ with the ratio of the areas of bipolar transistors result in

$$I_O = \frac{V_{th}}{R_1}\ln\left(\frac{A_2}{A_1}\right) = \frac{kT}{qR_1}\ln\left(\frac{A_2}{A_1}\right) = \left[\frac{k}{qR_1}\ln\left(\frac{A_2}{A_1}\right)\right]\times T. \tag{1.188}$$

The temperature coefficient of the output current will be

$$TC_{I_O} = \frac{dI_O}{dT} = \frac{k}{qR_1}\ln\left(\frac{A_2}{A_1}\right). \tag{1.189}$$

Setting $A_2 > A_1$, the output current I_O will linearly increase with temperature, so it will have a PTAT behavior.

1.6.2.1 Design Examples

In order to confirm the previously presented theoretical estimated results, the self-biased PTAT current source from Fig. 1.91 is simulated with LTspice using the following values: $K'_n = 200\,\mu\text{A}/\text{V}^2$, $K'_p = 100\,\mu\text{A}/\text{V}^2$, $\lambda_n = 0.01\,\text{V}^{-1}$, $\lambda_p = 0.02\,\text{V}^{-1}$, $V_{Tn} = 0.5\,\text{V}$, $V_{Tp} = -0.5\,\text{V}$ and $V_{DD} = 1.8\,\text{V}$. All transistors have the aspect ratios given by $\left(\frac{W}{L}\right) = \frac{5\,\mu\text{m}}{0.5\,\mu\text{m}} = 10$, the ratio of the areas of transistors Q2 and Q1 is $\frac{A_2}{A_1} = 10$, while resistors have the following values: $R_1 = 10\,\text{k}\Omega$ and $R_2 = 10\,\text{k}\Omega$.

The theoretical value of the temperature coefficient of the output current is

$$TC_{I_O} = \frac{25.9\,\text{mV}}{298\,\text{K}}\frac{1}{10\,\text{k}\Omega}\ln(10) \cong 20\,\text{nA/K}. \tag{1.190}$$

For evaluating the temperature behavior of the output current I_O with LTspice, a *DC sweep analysis* is performed, with the variation range for the temperature being set between 0 and 100 °C. The circuit used for this simulation is presented in Fig. 1.92, with the graphical representation of the temperature dependence of the output current being shown in Fig. 1.93. In order to directly view the result of the simulation, the directive ".save I(R2)" is introduced in the file, with I (R2) representing the output current of the circuit I_O.

The simulated value of the temperature coefficient of the output current is

$$TC_{I_O} = \frac{7.632\,\mu\text{A} - 5.532\,\mu\text{A}}{100\,\text{K}} = 21\,\text{nA/K}, \tag{1.191}$$

close to the theoretical value obtained using Eq. (1.190).

For evaluating the dependence of the output current I_O on the supply voltage variations with LTspice, a *DC sweep analysis* is performed, with the variation range

Fig. 1.92 Circuit used for simulating the dependence $I_O(T)$

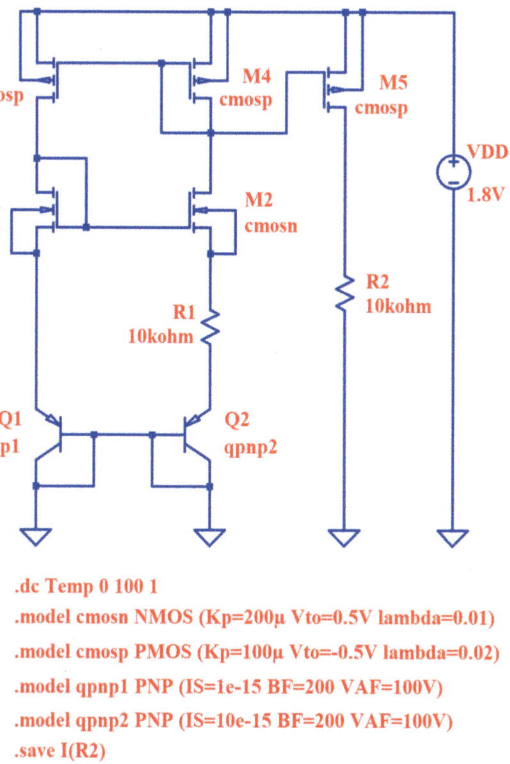

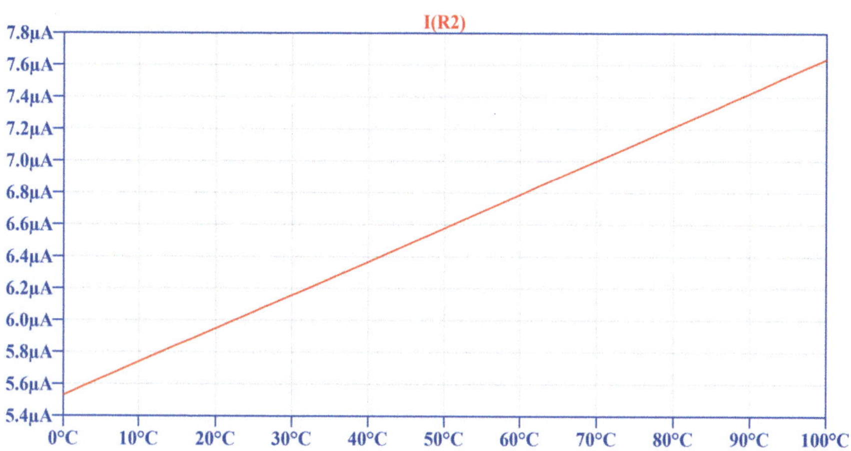

Fig. 1.93 The graphical representation of the dependence $I_O(T)$

Fig. 1.94 Circuit used for simulating the dependence $I_O(V_{DD})$

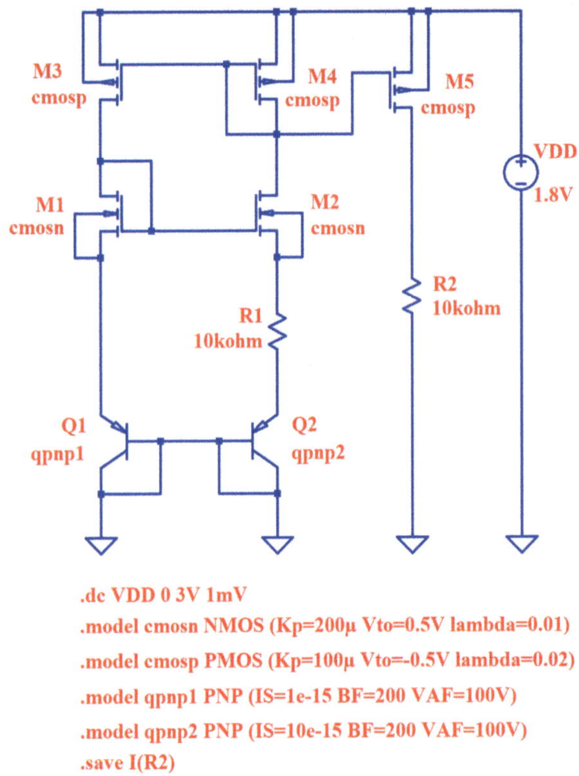

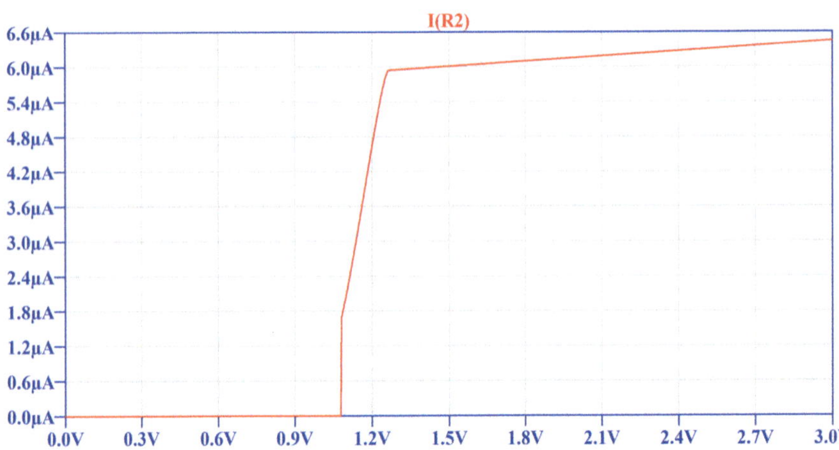

Fig. 1.95 Graphical representation of the dependence $I_O(V_{DD})$

for the supply voltage being set between 0 and 3 V. The circuit used for this simulation is presented in Fig. 1.94, with the graphical representation of the dependence $I_O(V_{DD})$ being shown in Fig. 1.95. In order to directly view the result of the simulation, the directive ".save I(R2)" is introduced in the file, with I (R2) representing the output current of the circuit I_O.

References

1. Razavi B (2001) Design of Analog CMOS Integrated Circuits. New York: McGraw-Hill Education
2. Gray PR, Hurst PJ, Lewis SH and Meyer RG (2009) Analysis and Design of Analog Integrated Circuits, 5th ed. New York: Wiley
3. Allen PE and Holberg DR (2011) CMOS Analog Circuit Design, 3rd ed. New York: Oxford University Press
4. Sedra AS and Smith KC (2015) Microelectronic Circuits, 7th ed. New York: Oxford University Press

Chapter 2
Analysis and Design of CMOS Differential Amplifiers: A Systematic Analysis and Design Approach

Abstract Differential amplifiers are fundamental building blocks in analog integrated circuit design. Their ability to amplify the difference between two input signals while rejecting common-mode components makes them essential in applications from signal conditioning to input stages of high-performance operational amplifiers. This chapter presents the analysis and design of CMOS differential amplifiers using a simplified transistor model. MOS devices are described by the quadratic current-voltage relationship in saturation, including channel-length modulation effect, ensuring a realistic yet analytically tractable framework. This approach enables closed-form derivations while maintaining good agreement with simulations. The chapter begins with the NMOS differential amplifier with passive load. Key parameters such as differential- and common-mode gain are derived using the half-circuit technique. Large-signal behavior, common-mode input range and input offset voltage are also examined. The analysis then extends to the PMOS differential amplifier, followed by a comparison of both structures. Finally, differential amplifiers with active loads, particularly current mirror configurations, are introduced, highlighting their higher gain and widespread use. LTspice simulations validate theoretical results and illustrate practical performance, providing a consistent framework for understanding operation, limitations and design trade-offs.

Keywords MOS differential amplifier · Active load · Differential-mode voltage gain · Common-mode voltage gain · Maximal range of the common-mode input voltage · Input offset voltage

2.1 Introduction

Differential amplifiers represent one of the fundamental building blocks in analog integrated circuit design and are widely documented in the literature [1–4]. Their ability to amplify the difference between two input signals while rejecting common-mode components makes them essential in a wide range of applications, from simple signal conditioning to input stages of high-performance operational amplifiers.

In this chapter, the analysis and design of CMOS differential amplifiers are developed using a simplified transistor model. The MOS transistors are described

C. R. Popa, *Fundamentals of CMOS Analog Integrated Circuits*,
https://doi.org/10.1007/978-3-032-30677-7_2

using the quadratic current-voltage relationship in saturation, complemented only by the channel-length modulation effect. This modelling approach enables fully analytical calculations while maintaining sufficient accuracy for meaningful comparison with circuit simulations.

The chapter begins with the study of the basic NMOS differential amplifier with passive load, where key performance metrics such as differential-mode gain and common-mode gain are derived using the half-circuit method. Additionally, the large signal transfer characteristic, the maximal range of the common-mode input voltage and the input offset voltage are also studied in detail. Next, the analysis of the basic PMOS differential amplifier with passive load is performed, and a detailed comparison between these complementary basic differential amplifiers is made, evidentiating the advantages and the limitations of each circuit. The analysis is then extended to differential amplifiers with active loads, including current mirror configurations, which significantly enhance gain and are widely used in integrated circuit design.

Throughout the chapter, theoretical results are systematically validated through LTspice simulations, highlighting the accuracy and limitations of the simplified analytical model.

All LTspice simulation files associated with the circuits presented in this book are provided as a supplementary material on the publisher's website. Each file is named identically to the corresponding figure in the chapter, with the extension ".asc". The reader is encouraged to run, analyse and modify these files in order to validate the theoretical results and explore circuit behavior. The use of these materials requires the installation of the LTspice simulation software.

2.2 Basic NMOS Differential Amplifier with Passive Load

The basic NMOS differential amplifier with passive load has a theoretically perfect symmetrical structure. It must ideally amplify only the differential-mode input voltage and must reject the common-mode input voltage.

The circuit of the NMOS differential amplifier is presented in Fig. 2.1. It consists of two identical NMOS transistors (M1 and M2) and two identical load resistances (R_1 and R_2). The NMOS differential amplifier with passive load is biased using a current source I_O, having an output resistance R_O.

2.2.1 *Large Signal Analysis*

In order to determine the capability of the circuit to amplify differential signals, the linearity range and the differential gain must be determined. These parameters of the differential amplifier could be obtained evaluating the large signal characteristic of

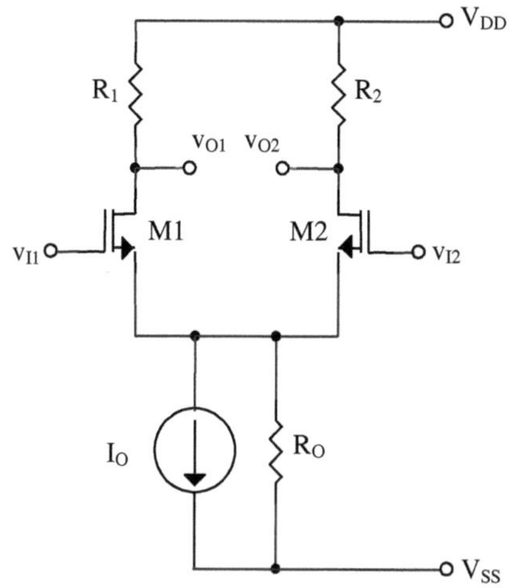

Fig. 2.1 Basic NMOS differential amplifier with passive load

the circuit, which is represented by the mathematical dependence of the differential-mode output voltage v_{OD} as a function of the differential-mode input voltage v_{ID}.

The large signal behavior of MOS differential amplifier is governed by the square-law device characteristics and is treated in detail in classical analog CMOS literature [1–4]. Because both transistors M1 and M2 are biased in saturation region, their drain currents can be expressed as functions of the gate-source voltages as follows:

$$i_{D1} = \frac{K'_n}{2}\left(\frac{W}{L}\right)(v_{GS1} - V_T)^2 \tag{2.1}$$

and, respectively,

$$i_{D2} = \frac{K'_n}{2}\left(\frac{W}{L}\right)(v_{GS2} - V_T)^2, \tag{2.2}$$

resulting in the complementary relations $v_{GS}(i_D)$:

$$v_{GS1} = V_T + \sqrt{\frac{2i_{i_{D1}}}{K'_n\left(\frac{W}{L}\right)}} \tag{2.3}$$

and

$$v_{GS2} = V_T + \sqrt{\frac{2i_{i_{D2}}}{K'_n\left(\frac{W}{L}\right)}}. \tag{2.4}$$

The channel-length modulation effect has been neglected for simplifying this mathematical analysis.

The differential-mode input voltage v_{ID} can be expressed as follows:

$$v_{ID} = v_{GS1} - v_{GS2}. \tag{2.5}$$

Replacing Eqs. (2.3) and (2.4) in Eq. (2.5) results in

$$v_{ID} = \sqrt{\frac{2}{K'_n\left(\frac{W}{L}\right)}}\left(\sqrt{i_{D1}} - \sqrt{i_{D2}}\right). \tag{2.6}$$

Considering a large value of the output resistance R_O of the current source that biases the differential amplifier, the sum of the drain currents is equal to the constant current I_O:

$$i_{D1} + i_{D2} = I_O \tag{2.7}$$

From Eqs. (2.5), (2.6) and (2.7), the following second-order equation having as unknown value the current i_{D1} results

$$i_{D1}^2 - I_O i_{D1} + \frac{1}{4}\left[I_O - \frac{K'_n}{2}\left(\frac{W}{L}\right)v_{ID}^2\right]^2 = 0. \tag{2.8}$$

Solving the previous equation, it is possible to obtain the mathematical dependencies of drain currents $i_{D1}(v_{ID})$ and $i_{D2}(v_{ID})$:

$$i_{D1} = \frac{I_O}{2} + \frac{I_O}{2}\sqrt{K'_n\left(\frac{W}{L}\right)\frac{v_{ID}^2}{I_O} - \frac{(K'_n)^2}{4}\left(\frac{W}{L}\right)^2\frac{v_{ID}^4}{I_O^2}} \tag{2.9}$$

and

$$i_{D2} = \frac{I_O}{2} - \frac{I_O}{2}\sqrt{K'_n\left(\frac{W}{L}\right)\frac{v_{ID}^2}{I_O} - \frac{(K'_n)^2}{4}\left(\frac{W}{L}\right)^2\frac{v_{ID}^4}{I_O^2}}. \tag{2.10}$$

When the first drain current i_{D1} increases, the second drain current i_{D2} decreases, with their sum remaining always constant and equal to the biasing current I_O. It is possible to determine the values of the differential-mode input voltage v_{ID} for which the entire biasing current I_O is computed in one branch, imposing the conditions that $i_{D1} = I_O$ and $i_{D2} = 0$. Using the notation $x \overset{not}{=} K'_n\left(\frac{W}{L}\right)\frac{v_{ID}^2}{I_O}$, the condition $i_{D1} = I_O$ becomes

$$\sqrt{x - \frac{x^2}{4}} = 1, \tag{2.11}$$

equivalent to

$$x^2 - 4x + 4 = 0 \tag{2.12}$$

resulting in $x = 2$. So, the values of the differential-mode input voltage v_{ID} for which ($i_{D1} = I_O$ and $i_{D2} = 0$) or ($i_{D1} = 0$ and $i_{D2} = I_O$) are, respectively,

$$(v_{ID})_{1,2} = \pm\sqrt{\frac{2I_O}{K'_n\left(\frac{W}{L}\right)}}. \tag{2.13}$$

The large signal characteristic of the NMOS differential amplifier with passive load can be obtained computing the differential-mode output voltage v_{OD}:

$$v_{OD} = v_{O1} - v_{O2} = R_1(i_{D2} - i_{D1}) \tag{2.14}$$

(because the load resistances R_1 and R_2 are identical, $R_1 = R_2 \stackrel{not}{=} R$). Replacing Eqs. (2.9) and (2.10) in Eq. (2.14) results in

$$v_{OD} = -\frac{Rv_{ID}}{2}\sqrt{4K'_n\left(\frac{W}{L}\right)I_O - (K'_n)^2\left(\frac{W}{L}\right)^2 v_{ID}^2}. \tag{2.15}$$

The transfer characteristic $v_{OD}(v_{ID})$ is strongly nonlinear, so the circuit presented in Fig. 2.1 can be used for amplifying signals only for a restricted range of the differential-mode input voltage v_{ID}. The decision of how much must the range of variation be limited for the input voltage is correlated with the maximal distortions that could be accepted in the process of amplifying the input signal, as it is shown in the following distortion analysis.

In order to quantitatively evaluate the nonlinearity of the transfer characteristic for the basic NMOS differential amplifier with passive load, the mathematical function $v_{OD}(v_{ID})$ must be developed in Taylor series as follows:

$$v_{OD}(v_{ID}) = a_1 v_{ID} + a_3 v_{ID}^3 + a_5 v_{ID}^5 + \ldots \tag{2.16}$$

In consequence, the mathematical function $v_{OD}(v_{ID})$ can be expressed as a polynomial function, having only odd-order terms. This particular behavior of the circuit is a result of its perfect theoretical symmetry. Coefficients a_1, a_3, a_5, ... can be obtained from Eq. (2.15) by determining the superior-order derivatives of $v_{OD}(v_{ID})$ function, computed for $v_{ID} = 0$:

$$a_1 = -R\left[K'_n\left(\frac{W}{L}\right)I_O\right]^{1/2}, \tag{2.17}$$

$$a_3 = \frac{R}{8I_O^{1/2}}\left[K'_n\left(\frac{W}{L}\right)\right]^{3/2}, \tag{2.18}$$

$$a_5 = -\frac{R}{128I_O^{3/2}}\left[K'_n\left(\frac{W}{L}\right)\right]^{5/2}. \tag{2.19}$$

In Eq. (2.16), the a_1 coefficient represents the differential-mode voltage gain of the circuit presented in Fig. 2.1:

$$A_{dd} = a_1 = -R\sqrt{K'_n\left(\frac{W}{L}\right)I_O} \tag{2.20}$$

All superior-order odd terms from the same relation (a_k, with $k \geq 3$) represent the mathematical modelling of the distortions introduced by the circuit. In order to be able to use this circuit for amplifying signals with reasonable distortions, all these superior-order terms must be neglectable in comparison with the useful linear term, $a_1 v_{ID}$. Because the most important distortion is introduced by the third-order term from Eq. (2.16), the total order distortion coefficient of the differential amplifier is approximatively given by the ratio between the third-order term and the linear term from Eq. (2.16):

$$THD \cong THD_3 = \left|\frac{a_3 v_{ID}^3}{a_1 v_{ID}}\right|, \tag{2.21}$$

equivalent to

$$THD \cong \frac{v_{ID}^2}{8I_O} K'_n\left(\frac{W}{L}\right). \tag{2.22}$$

So, the total order distortion coefficient *THD* of the differential amplifier is proportional to the square of the differential-mode input voltage v_{ID}. In consequence, the maximal value, which is acceptable for the differential-mode input voltage, is strictly correlated to the maximal value of the total order distortion coefficient, considered to be acceptable for a particular application. Usually, an order of magnitude of some per cent could be considered to be reasonable for most applications.

2.2.2 Small Signal Analysis

The small signal analysis of MOS differential amplifier is based on well-established analog circuit theory [1–4]. Considering a limited differential-mode signal applied to the input of the differential amplifier presented in Fig. 2.1, with a maximal value imposed by the condition of signal amplifying with reasonable distortions, the behavior of the circuit becomes approximately linear. In this case, the previous complicated large signal analysis can be replaced with a very simple small signal analysis. In order to better understand the behavior of the differential amplifier, the differential-mode and common-mode voltages must be rigorously defined.

Using the notations v_{I1} and v_{I2} for the input voltages of the basic NMOS differential amplifier with passive load presented in Fig. 2.1 and the notations v_{O1} and v_{O2} for its output voltages, the differential-mode input voltage v_{ID} is defined as the difference between the input voltages:

$$v_{ID} = v_{I1} - v_{I2}, \tag{2.23}$$

and the common-mode input voltage v_{IC} is defined as the arithmetical mean of the input voltages:

$$v_{IC} = \frac{v_{I1} + v_{I2}}{2} \tag{2.24}$$

Similarly, the differential-mode and the common-mode output voltages v_{OD} and v_{OC} are defined as follows:

$$v_{OD} = v_{O1} - v_{O2} \tag{2.25}$$

and

$$v_{OC} = \frac{v_{O1} + v_{O2}}{2} \tag{2.26}$$

From another perspective, it could be considered that the double information related to the input (voltages v_{I1} and v_{I2}) is perfectly equivalent with another double information (voltages v_{ID} and v_{IC}). Knowing voltages v_{ID} and v_{IC}, it is possible to determine from Eqs. (2.23) and (2.24) the voltages v_{I1} and v_{I2}, as follows:

$$v_{I1} = v_{IC} + \frac{v_{ID}}{2} \tag{2.27}$$

and

$$v_{I2} = v_{IC} - \frac{v_{ID}}{2} \tag{2.28}$$

Similarly, it could also be considered that the double information related to the output (voltages v_{O1} and v_{O2}) is perfectly equivalent with another double information (voltages v_{OD} and v_{OC}). Knowing voltages v_{OD} and v_{OC}, it is possible to determine from Eqs. (2.25) and (2.26) the voltages v_{O1} and v_{O2}, as follows:

$$v_{O1} = v_{OC} + \frac{v_{OD}}{2} \tag{2.29}$$

and

$$v_{O2} = v_{OC} - \frac{v_{OD}}{2}. \tag{2.30}$$

In conclusion, using the differential-mode and common-mode voltages (for both input and output), it is possible to have in an equivalent way the double information related to the input of the differential amplifier and also the double information related to its output. From a practical perspective, this alternative mathematical approach presents the very important advantage of a much better understanding of the general behavior of any differential amplifier (not only of the basic NMOS differential amplifier with passive load):

- A differential amplifier must ***amplify*** as much as possible the differential-mode input voltage.
- A differential amplifier must ***reject*** as much as possible the common-mode input voltage.

In order to quantitatively evaluate this double behavior, it is useful to define four voltage gains, specific to any differential amplifier. There exist four voltage gains because the differential amplifier has two input voltages (v_{I1} and v_{I2}) and two output voltages (v_{O1} and v_{O2}).

(a) *The differential-mode input/differential-mode output voltage gain* A_{DD}:

$$A_{DD} \overset{def}{=} \left. \frac{v_{OD}}{v_{ID}} \right|_{v_{IC}=0} \tag{2.31}$$

(b) *The common-mode input/common-mode output voltage gain* A_{CC}:

$$A_{CC} \overset{def}{=} \left. \frac{v_{OC}}{v_{IC}} \right|_{v_{ID}=0} \tag{2.32}$$

(c) *The common-mode input/differential-mode output voltage gain* A_{CD}:

$$A_{CD} \overset{def}{=} \left. \frac{v_{OD}}{v_{IC}} \right|_{v_{ID}=0} \tag{2.33}$$

(d) The differential-mode input/common-mode output voltage gain A_{CD}:

$$A_{DC} \stackrel{def}{=} \left. \frac{v_{OC}}{v_{ID}} \right|_{v_{IC}=0} \tag{2.34}$$

Using these definitions of the voltage gains for the differential amplifier, the differential-mode and common-mode output voltages can be expressed as follows:

$$v_{OD} = A_{DD} v_{ID} + A_{CD} v_{IC} \tag{2.35}$$

and

$$v_{OC} = A_{DC} v_{ID} + A_{CC} v_{IC}, \tag{2.36}$$

resulting in the following expressions of the output voltages of the differential amplifier:

$$v_{O1} = \left(A_{DC} + \frac{A_{DD}}{2}\right) v_{ID} + \left(A_{CC} + \frac{A_{CD}}{2}\right) v_{IC} \tag{2.37}$$

and

$$v_{O2} = \left(A_{DC} - \frac{A_{DD}}{2}\right) v_{ID} + \left(A_{CC} - \frac{A_{CD}}{2}\right) v_{IC}. \tag{2.38}$$

A very important advantage of the differential amplifier is that, as a consequence of its symmetry, the voltage gains A_{DC} and A_{CD} are equal to zero. So, the previous relations become

$$v_{O1} = \frac{A_{DD}}{2} v_{ID} + A_{CC} v_{IC}, \tag{2.39}$$

and

$$v_{O2} = -\frac{A_{DD}}{2} v_{ID} + A_{CC} v_{IC}. \tag{2.40}$$

The previous two relations represent the mathematical basis for a very important conclusion related to the utilization of the basic NMOS differential amplifier with passive load. In order to detail this conclusion, two hypotheses must be considered:

- The utilization of the voltage $v_{OD} = v_{O1} - v_{O2}$ as the output voltage of the differential amplifier (differential-mode output voltage): In this case, because $v_{OD} = v_{O1} - v_{O2} = A_{DD}v_{ID}$, the differential-mode voltage gain of the circuit will be equal to A_{DD} (see Eqs. (2.39) and (2.40)).
- The utilization of either voltage v_{O1} or voltage v_{O2} as the output voltage of the differential amplifier (single-mode output voltage): In this case, the differential-mode voltage gain of the circuit will be equal only to $\frac{A_{DD}}{2}$ (see Eqs. (2.39) and (2.40)).

2.2.2.1 Computation of the Differential-Mode Voltage Gain A_{DD} Using the Half-Circuit Method

Because of the symmetry of the basic NMOS differential amplifier with passive load, the half-circuit method could be used for computing the differential-mode voltage gain A_{DD}. A full differential operation of the differential amplifier is illustrated in Fig. 2.2.

The circuit receives as input voltages only differential-mode voltages, that is, $v_{I1} = \frac{v_{ID}}{2}$ and $v_{I2} = -\frac{v_{ID}}{2}$ (Eqs. (2.27) and (2.28), particularized for common-mode input voltage $v_{IC} = 0$). Similarly, the circuit generates only differential-mode voltages, that is, $v_{O1} = \frac{v_{OD}}{2}$ and $v_{O2} = -\frac{v_{OD}}{2}$ (Eqs. (2.29) and (2.30), particularized for common-mode output voltage $v_{OC} = 0$). The supply voltages V_{DD} and V_{SS} and also the current source I_O are set to zero for the circuit presented in Fig. 2.1.

If a differential-mode output voltage ($v_{OD} = v_{O1} - v_{O2}$) is considered, the differential-mode voltage gain of the circuit presented in Fig. 2.2 will be defined as follows:

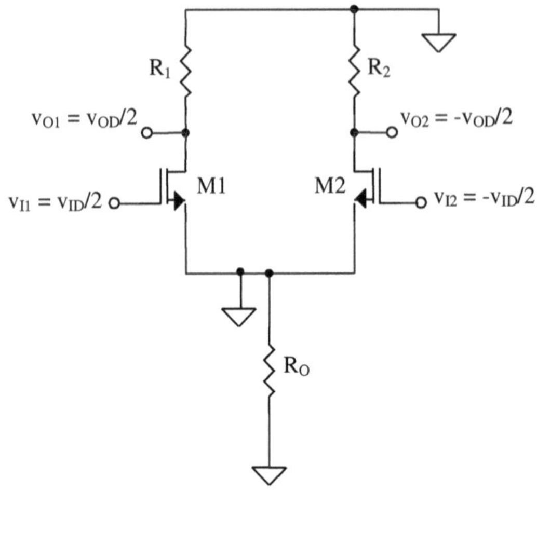

Fig. 2.2 The circuit used for evaluating the differential-mode voltage gain A_{DD} using the half-circuit method

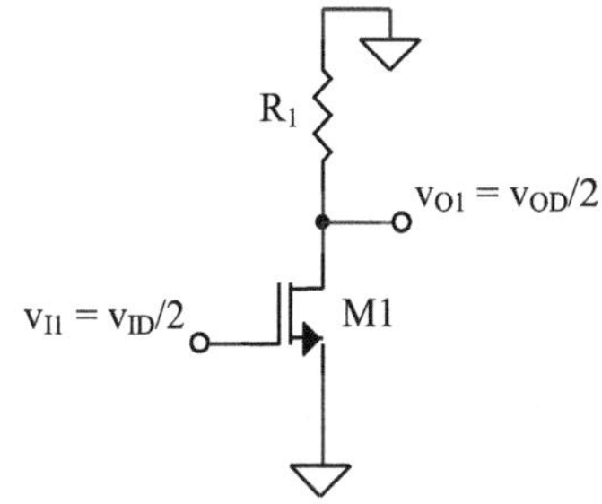

Fig. 2.3 Differential-mode half circuit for the basic NMOS differential amplifier with passive load

$$A_{DD} = \frac{v_{O1} - v_{O2}}{v_{I1} - v_{I2}} = \frac{\frac{v_{OD}}{2} - \left(-\frac{v_{OD}}{2}\right)}{\frac{v_{ID}}{2} - \left(-\frac{v_{ID}}{2}\right)} = \frac{v_{OD}}{v_{ID}}. \tag{2.41}$$

If a single-mode output voltage (v_{O1} or v_{O2}) is considered, the differential-mode voltage gain of the circuit presented in Fig. 2.2 will be defined as follows:

$$A_{DD} = \frac{v_{O1}}{v_{I1} - v_{I2}} = \frac{\frac{v_{OD}}{2}}{\frac{v_{ID}}{2} - \left(-\frac{v_{ID}}{2}\right)} = \frac{1}{2}\frac{v_{OD}}{v_{ID}}. \tag{2.42}$$

Based on the symmetry of the previous circuit and knowing the information that the potential in the point of connection between the source of transistor M1 and the source of transistor M2 is zero (virtual ground point), the differential-mode half circuit of the differential amplifier presented in Fig. 2.2 is shown in Fig. 2.3.

The previous half circuit represents a basic common-source amplifier, having the following expression of the voltage gain A_{DD-HC}:

$$A_{DD-HC} = \frac{\frac{v_{OD}}{2}}{\frac{v_{ID}}{2}} = \frac{v_{OD}}{v_{ID}} = -g_{m1}(R_1//r_{ds1}). \tag{2.43}$$

Comparing Eqs. (2.41), (2.42) and (2.43), the final conclusion related to the utilization of the NMOS differential-mode half circuit is the following:

- For the utilization of the voltage $v_{O1} - v_{O2} = v_{OD}$ as the output voltage of the differential amplifier (differential-mode output voltage), the differential-mode voltage gain of the basic NMOS differential amplifier with passive load is equal to the voltage gain of the differential-mode half circuit, $A_{DD} = A_{DD-HC} = -g_{m1}(R_1//r_{ds1})$.
- For the utilization of the voltage $v_{O1} = \frac{v_{OD}}{2}$ or of the voltage $v_{O2} = -\frac{v_{OD}}{2}$ as the output voltage of the differential amplifier (single-mode output voltage), the differential-mode voltage gain of the basic NMOS differential amplifier with passive load is equal to half of the voltage gain of the differential-mode half circuit, $A_{DD} = \frac{A_{DD-HC}}{2} = -\frac{1}{2}g_{m1}(R_1//r_{ds1})$.

2.2.2.2 Computation of the Common-Mode Voltage Gain A_{CC} Using the Half-Circuit Method

Because of the symmetry of the basic NMOS differential amplifier with passive load, the half-circuit method could also be used for computing the common-mode voltage gain A_{CC}. An exclusive common-mode operation of the differential amplifier is illustrated in Fig. 2.4.

The circuit receives as input voltages only common-mode voltages, that is, $v_{I1} = v_{IC}$ and $v_{I2} = v_{IC}$ (Eqs. (2.27) and (2.28), particularized for differential-mode input voltage $v_{ID} = 0$). Similarly, the circuit generates only common-mode voltages, that is, $v_{O1} = v_{OC}$ and $v_{O2} = v_{OC}$ (Eqs. (2.29) and (2.30), particularized for differential-mode output voltage $v_{OD} = 0$). The supply voltages V_{DD} and V_{SS} and also the current source I_O are set to zero for the circuit presented in Fig. 2.1. The common-mode voltage gain can be defined as follows:

$$A_{CC} = \frac{v_{OC}}{v_{IC}}. \tag{2.44}$$

Based on the symmetry of the previous circuit, the common-mode half circuit of the differential amplifier presented in Fig. 2.1 is shown in Fig. 2.5. For obtaining a perfect symmetrical circuit from the differential amplifier shown in Fig. 2.4, the single resistor R_O can be virtually replaced with a parallel connection of two identical resistors, both equal to $2R_O$.

The previous half circuit represents a basic source-degenerated amplifier, having the following expression of the voltage gain A_{CC-HC}:

$$A_{CC-HC} = \frac{v_{OC}}{v_{IC}} = -\frac{g_{m1}R_1}{1 + 2g_{m1}R_O} \cong -\frac{R_1}{2R_O}. \tag{2.45}$$

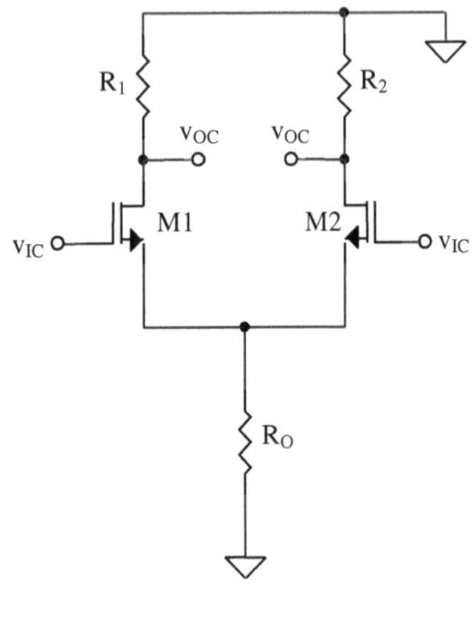

Fig. 2.4 The circuit used for evaluating the common-mode voltage gain A_{CC} using the half-circuit method

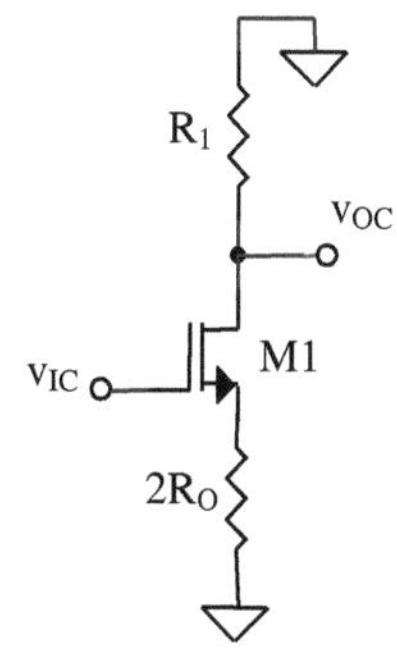

Fig. 2.5 Common-mode half circuit for the basic NMOS differential amplifier with passive load

Comparing Eqs. (2.44) and (2.45), it results that the common-mode voltage gain of the basic NMOS differential amplifier with passive load is equal to the voltage gain of the common-mode half-circuit, $A_{CC} = A_{CC\ -\ HC}$. So,

$$A_{CC} \cong -\frac{R_1}{2R_O}. \tag{2.46}$$

It is possible to define the common-mode rejection ratio (CMRR) for the basic NMOS differential amplifier with passive load as follows:

$$\text{CMRR} = \left|\frac{A_{DD}}{A_{CC}}\right|. \tag{2.47}$$

Using Eqs. (2.43) and (2.46) for obtaining the expressions of A_{DD} and A_{CC} and replacing them in Eq. (2.47) result in

$$\text{CMRR} \cong 2g_{m1}R_O. \tag{2.48}$$

In Eq. (2.43), the value of resistance r_{ds1} was neglected in comparison to resistance R_1.

In order to increase CMRR, it is necessary to decrease the absolute value of the common-mode voltage gain A_{CC}. The most convenient method is the increasing of the output resistance R_O of the current source that biases the basic NMOS differential amplifier with passive load. In consequence, the biasing of the differential amplifier using a cascode current source will strongly decrease the common-mode voltage gain of the differential amplifier. Unfortunately, this decision comes with at least two disadvantages:

- The increasing of the minimal supply voltage of the differential amplifier.
- The limitation of the maximal range of the common-mode input voltage for the differential amplifier (a parameter that is evaluated in the following paragraph).

2.2.3 Maximal Range of the Common-Mode Input Voltage

A differential amplifier must amplify as much as possible the differential-mode input voltage v_{ID} and must reject as much as possible the common-mode input voltage V_{IC}. Another problem related to the common-mode input voltage is the maximal range of this voltage $\left[V_{IC}^{MIN}, V_{IC}^{MAX}\right]$. If the common-mode voltage applied on the inputs of a differential amplifier is superior to the maximal acceptable value V_{IC}^{MAX}, or inferior to the minimal acceptable value V_{IC}^{MIN}, a part of the circuit does no longer work and all the differential amplifier operation is compromised. For this reason, it is extremely important to determine the theoretical limits of this maximal range of the common-mode input voltage and to use the differential amplifier only for common-mode input voltages that are inside this interval.

In order to determine the maximal range of the common-mode input voltage for the basic NMOS differential amplifier with passive load, it is necessary to choose a particular implementation of the current source I_O for biasing the differential amplifier presented in Fig. 2.1. Because the most used current sources from Chap. 1 are represented by the MOS basic current mirror from Fig. 1.16, this current source is used in Fig. 2.6 for biasing the basic NMOS differential amplifier with passive load.

The maximal value of the common-mode input voltage is limited by the operation in saturation of both transistors M1 and M2:

$$V_{IC}^{MAX} = V_{DD} - \frac{I_O}{2}R_1 - (V_{GS1} - V_T) + V_{GS1} = V_{DD} - \frac{I_O}{2}R_1 + V_T. \quad (2.49)$$

The increasing of V_{IC}^{MAX} is possible by decreasing the load resistance R_1 or by decreasing the biasing current I_O. Unfortunately, both methods will decrease the differential-mode voltage gain A_{DD}. Sacrificing this gain, it is theoretically possible to obtain $V_{IC}^{MAX} > V_{DD}$, having the significance that the differential amplifier is

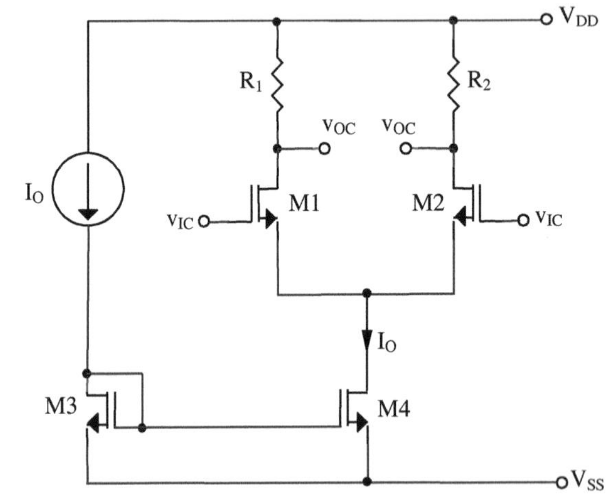

Fig. 2.6 Basic NMOS differential amplifier with passive load, biased using a basic MOS current mirror

extremely tolerant: it can accept a common-mode input voltage superior to the value of its positive supply voltage (of course, this voltage can be applied at their inputs only by another circuit, which is supplied at a voltage superior to V_{DD}).

The minimal value of the common-mode input voltage is limited by the operation in saturation of transistor M4:

$$V_{IC}^{\mathrm{MIN}} = V_{SS} + (V_{GS4} - V_T) + V_{GS1} \tag{2.50}$$

Using the approximate squaring characteristic of transistors M1 and M4, biased in saturation region, the previous relation becomes

$$V_{IC}^{\mathrm{MIN}} = V_{SS} + V_T + \left(\sqrt{2} + 1\right)\sqrt{\frac{I_O}{K'_n\left(\frac{W}{L}\right)}}. \tag{2.51}$$

The only method for improving the behavior of the basic NMOS differential amplifier with passive load from the perspective of its minimal common-mode input voltage is to decrease the biasing current I_O. The disadvantage of this approach is the reducing of the differential-mode voltage gain A_{DD}, because of the reduction of transconductances g_{m1} and g_{m2} of transistors M1 and M2. Even accepting this disadvantage, the value of V_{IC}^{MIN} is still strongly limited in comparison to the value of V_{IC}^{MAX}.

In conclusion, the basic NMOS differential amplifier with passive load can be usually used especially for relatively large values of the common-mode input voltages. Applications that need a proper operation of the differential amplifier for small values of the common-mode input voltage must replace the basic NMOS differential amplifier with passive load with its complementary circuit, the basic PMOS differential amplifier with passive load.

2.2.3.1 Rail-to-Rail Differential Amplifier

Low-voltage applications that are able to operate at extremely low voltages (for example, $\pm 1V$) must use a rail-to-rail differential amplifier (an improved performance circuit that can correctly operate for an extended range of the common-mode input voltage V_{IC}, between $V_{SS} = -V_{DD}$ and V_{DD}). In order to obtain this behavior, the classical method is to use two complementary differential amplifiers. The basic NMOS differential amplifier with passive load will work excellent for large values of V_{IC}, while the basic PMOS differential amplifier with passive load will work excellent for small values of V_{IC}. The simplest method for connecting these complementary differential amplifiers is to use a parallel connection. The disadvantage is the obtaining of a double differential-mode voltage gain for medium values of V_{IC}, because in this area, both differential amplifiers are properly working. A better method for avoiding this undesired effect is to use circuits that are able to extract the maximal value between two currents or to use transconductance computational circuits. In this case, the variation between the different values of the differential-mode voltage gains when V_{IC} varies between $V_{SS} = -V_{DD}$ and V_{DD} is given only by the difference between gain capabilities of NMOS and PMOS differential amplifiers (for the first method) or this variation can be cancelled (for the second method), respectively.

2.2.4 *Input Offset Voltage*

The basic NMOS differential amplifier with passive load is, theoretical, a perfect symmetric structure. Nevertheless, in practice, there exist undesired asymmetries between the left branch and the right branch of the circuit. For example, the following parameters contain small asymmetries: threshold voltage V_T and aspect ratios W/L for transistors M1 and M2 or values of resistors R_1 and R_2. In consequence, the theoretical relation $v_{OD} = A_{DD}v_{ID}$ (obtained from Eq. (2.31)) changes, becoming, in the real situation of implementing the differential amplifier in a particular CMOS technology,

$$v_{OD} = A_{DD}v_{ID} + C, \tag{2.52}$$

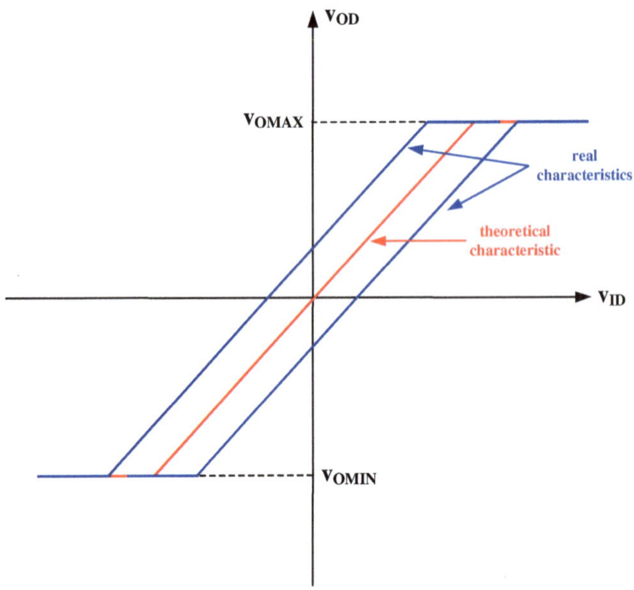

Fig. 2.7 Transfer characteristics for the basic NMOS passive load differential amplifier (theoretical circuit and real circuit, containing asymmetries)

with C being a constant that models the cumulus of all asymmetries. In Fig. 2.7, the following transfer characteristics are represented:

- The theoretical characteristic, which contains the origin point.
- Real characteristics, shifted to the left or to the right, depending on the absolute value of the cumulated asymmetries as well as the relationship that exists between the left branch and the right branch of the differential amplifier.

The strange behavior of the real differential amplifier (developed in any CMOS integrated technology using inherent technological asymmetric branches) will be a non-zero output voltage, obtained for a zero input voltage. The transfer characteristics for the basic NMOS passive load differential amplifier (theoretical circuit and real circuit, containing asymmetries) are presented in Fig. 2.7.

The input offset voltage v_{OS} quantitative evaluates the cumulated effect of these technological asymmetries, being defined as the voltage that must be applied at the input of an asymmetric differential amplifier in order to obtain a zero output voltage.

In order to determine the mathematical approximate expression of the input offset voltage, three asymmetries will be considered:

- Different threshold voltages V_{T1} and V_{T2} for transistors M1 and M2, representing the core of the differential amplifier:

$$V_{T1} = V_T \pm \frac{\Delta V_T}{2} \tag{2.53}$$

and

$$V_{T2} = V_T \mp \frac{\Delta V_T}{2} \tag{2.54}$$

- Different aspect ratios $(W/L)_1$ and $(W/L)_2$ for the same transistors:

$$\left(\frac{W}{L}\right)_1 = \left(\frac{W}{L}\right) \pm \frac{\Delta\left(\frac{W}{L}\right)}{2} \tag{2.55}$$

and

$$\left(\frac{W}{L}\right)_2 = \left(\frac{W}{L}\right) \mp \frac{\Delta\left(\frac{W}{L}\right)}{2} \tag{2.56}$$

- Different load resistors, R_1 and R_2:

$$R_1 = R \pm \frac{\Delta R}{2} \tag{2.57}$$

and

$$R_2 = R \mp \frac{\Delta R}{2}, \tag{2.58}$$

where ΔV_T, $\Delta\left(\frac{W}{L}\right)$ and ΔR represent the variations of V_T, $\left(\frac{W}{L}\right)$ and R, respectively, as a consequence of technological asymmetries. In order to simplify the mathematical form of the input offset voltage v_{OS}, the usual particularity of these asymmetries can be considered: their order of magnitudes is extremely small, about some per cent. This information can be mathematically modelled by the following approximations that can be used in the computation of the input offset voltage v_{OS}:

$$\frac{\Delta V_T}{V_T} << 1, \tag{2.59}$$

$$\frac{\Delta\left(\frac{W}{L}\right)}{\left(\frac{W}{L}\right)} << 1, \tag{2.60}$$

$$\frac{\Delta R}{R} << 1. \tag{2.61}$$

The expression of voltage v_{OS} can be written as follows:

$$v_{OS} = v_{GS1} - v_{GS2}. \tag{2.62}$$

Considering the approximate square-root dependence of the gate-source voltage on the drain current, the previous relation becomes

$$v_{OS} = (V_{T1} - V_{T2}) + \left(\sqrt{\frac{2i_{D1}}{K'_n\left(\frac{W}{L}\right)_1}} - \sqrt{\frac{2i_{D2}}{K'_n\left(\frac{W}{L}\right)_2}} \right), \tag{2.63}$$

equivalent to

$$v_{OS} = \Delta V_T + \sqrt{\frac{2\left(i_D + \frac{\Delta i_D}{2}\right)}{K'_n\left[\left(\frac{W}{L}\right) - \frac{1}{2}\Delta\left(\frac{W}{L}\right)\right]}} - \sqrt{\frac{2\left(i_D - \frac{\Delta i_D}{2}\right)}{K'_n\left[\left(\frac{W}{L}\right) + \frac{1}{2}\Delta\left(\frac{W}{L}\right)\right]}}, \tag{2.64}$$

or

$$v_{OS} \cong \Delta V_T + \sqrt{\frac{2i_D}{K'_n\left(\frac{W}{L}\right)}} \left[\sqrt{1 + \frac{\Delta i_D}{2i_D} + \frac{\Delta\left(\frac{W}{L}\right)}{2\left(\frac{W}{L}\right)}} - \sqrt{1 - \frac{\Delta i_D}{2i_D} - \frac{\Delta\left(\frac{W}{L}\right)}{2\left(\frac{W}{L}\right)}} \right]. \tag{2.65}$$

The definition of the input offset voltage v_{OS} imposes the condition that $v_{OD} = 0$. For this reason, the voltages across resistors R_1 and R_2 are equal, so

$$\left(i_D + \frac{\Delta i_D}{2}\right)\left(R - \frac{\Delta R}{2}\right) = \left(i_D - \frac{\Delta i_D}{2}\right)\left(R + \frac{\Delta R}{2}\right), \tag{2.66}$$

equivalent to

$$\frac{\Delta i_D}{i_D} = \frac{\Delta R}{R}. \tag{2.67}$$

Replacing Eq. (2.67) in Eq. (2.65) results in

$$v_{OS} \cong \Delta V_T + \sqrt{\frac{2i_D}{K'_n\left(\frac{W}{L}\right)}}\left[\sqrt{1 + \frac{\Delta R}{2R} + \frac{\Delta\left(\frac{W}{L}\right)}{2\left(\frac{W}{L}\right)}} - \sqrt{1 - \frac{\Delta R}{2R} - \frac{\Delta\left(\frac{W}{L}\right)}{2\left(\frac{W}{L}\right)}}\right]. \tag{2.68}$$

Using Eqs. (2.60) and (2.61), the previous relation can be further approximated using its limited Taylor series expansion, resulting in

$$v_{OS} \cong \Delta V_T + \frac{v_{GS} - V_T}{2}\left[\frac{\Delta R}{R} + \frac{\Delta\left(\frac{W}{L}\right)}{\left(\frac{W}{L}\right)}\right]. \tag{2.69}$$

So, the value of the input offset voltage v_{OS} is strongly affected by technologically caused errors and also by temperature variations. In consequence, it is possible only to partially compensate the errors introduced by the input offset voltage. This compensation can be made only for a limited range of the temperature variations, using circuit techniques with reasonable complexities.

2.2.5 *Design Examples*

In order to confirm the previously presented theoretically estimated results, the basic NMOS passive load differential amplifier from Fig. 2.6 is simulated with LTspice using the following values: $K'_n = 200\ \mu\text{A/V}^2$, $\lambda_n = 0.01\ \text{V}^{-1}$, $V_{Tn} = 0.5$ V, $V_{DD} = 1.8$ V and $V_{SS} = -1.8$V. MOS transistors have the aspect ratio $\left(\frac{W}{L}\right)_1 = \left(\frac{W}{L}\right)_2 = \left(\frac{W}{L}\right)_3 = \left(\frac{W}{L}\right)_4 = \frac{5\ \mu\text{m}}{0.5\ \mu\text{m}} = 10$. The values of resistors are $R_1 = R_2 = 100\ \text{k}\Omega$, while the value of the current source is $I_O = 20\ \mu\text{A}$.

Equation (2.13) can be used for determining the values of the differential-mode input voltage for complete current switching in one branch of the differential amplifier (left or right):

$$(v_{ID})_{1,2} = \pm\sqrt{\frac{2 \times 20\ \mu\text{A}}{200\ \mu\text{A/V}^2 \times 10}} \cong \pm 0.141\,\text{V} \tag{2.70}$$

In order to determine the theoretical value of the differential-mode voltage gain expressed using Eqs. (2.41) and (2.43), it is necessary to compute the transconductance g_{m1} and the internal resistance r_{ds1} of transistor M1:

$$g_{m1} = \sqrt{2 \times 200\,\mu\text{A/V}^2 \times 10 \times 10\,\mu\text{A}} = 200\,\mu\text{A/V}, \tag{2.71}$$

$$r_{ds1} = \frac{1}{0.01\,\text{V}^{-1} \times 10\,\mu\text{A}} = 10\,\text{M}\Omega. \tag{2.72}$$

The theoretical value of A_{DD} is

$$A_{DD} = -200\,\mu\text{A/V} \times (100\,\text{k}\Omega / 10\,\text{M}\Omega) \cong -20. \tag{2.73}$$

The theoretical value of the common-mode voltage gain A_{CC} can be computed using Eq. (2.46) and replacing the output resistance R_O of the current mirror M3–M4 with r_{ds4}:

$$A_{CC} = -\frac{100\,\text{k}\Omega}{2 \times \frac{1}{0.01\,\text{V}^{-1} \times 20\,\mu\text{A}}} = -\frac{100\,\text{k}\Omega}{10\,\text{M}\Omega} = -0.01. \tag{2.74}$$

The theoretical value of the maximal value of the common-mode input voltage is expressed by Eq. (2.49):

$$V_{IC}^{MAX} = 1.8\,\text{V} - \frac{20\,\mu\text{A}}{2} \times 100\,\text{k}\Omega + 0.5\,\text{V} = 1.3\text{V}. \tag{2.75}$$

The theoretical value of the minimal common-mode input voltage is expressed by Eq. (2.51):

$$V_{IC}^{MIN} = -1.8\,\text{V} + 0.5\,\text{V} + \left(\sqrt{2} + 1\right)\sqrt{\frac{20\,\mu\text{A}}{200\,\mu\text{A/V}^2 \times 10}} \cong -1.059\,\text{V}. \tag{2.76}$$

2.2.5.1 Large Signal Operation Simulations

In order to graphically represent the dependencies of the drain currents of transistors M1 and M2 on the differential-mode input voltage v_{ID}, for the basic NMOS passive load differential amplifier from Fig. 2.6, a *DC sweep analysis* is performed in LTspice, for a variation of the voltage v_{ID} between $-200\,\text{mV}$ and $200\,\text{mV}$. The

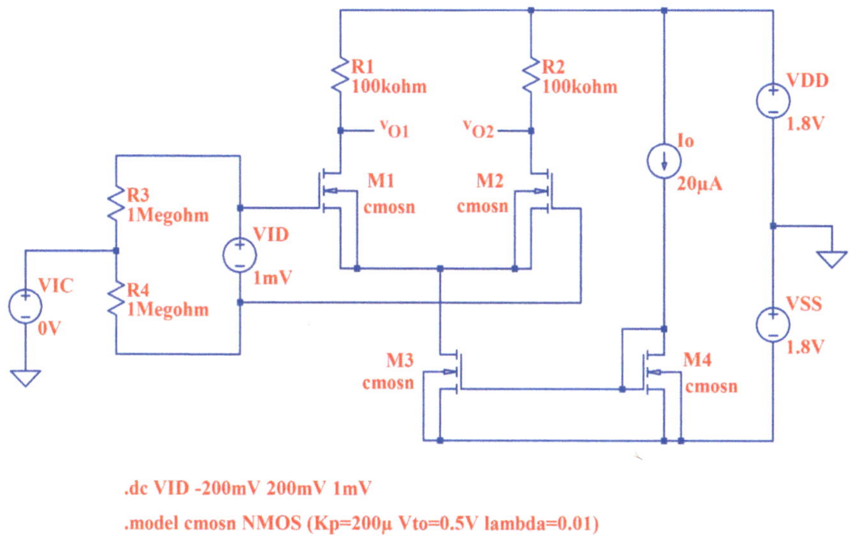

Fig. 2.8 The circuit used for simulating the dependencies $i_{D1}(v_{ID})$ and $i_{D2}(v_{ID})$

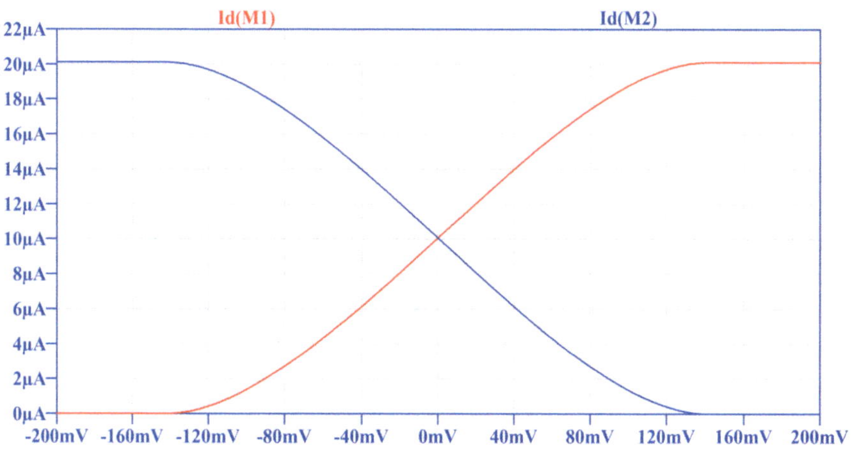

Fig. 2.9 The dependencies $i_{D1}(v_{ID})$ and $i_{D2}(v_{ID})$

circuit used for this simulation is presented in Fig. 2.8 (it is necessary to set two current markers in LTspice, in the drains of transistors M1 and M2), and the simulation results are shown in Fig. 2.9.

The simulated value of the differential-mode input voltage for complete current switching in one branch of the differential amplifier (left or right) is approximately equal to 0.141 V, identical to the theoretical value obtained in Eq. (2.70).

For illustrating the influence of the biasing current I_O on the operation of the differential amplifier presented in Fig. 2.10, a *DC sweep analysis* is performed in

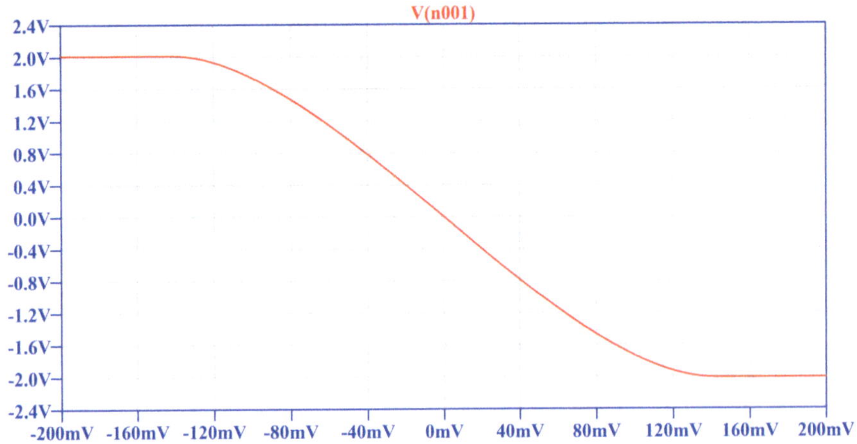

Fig. 2.13 The transfer characteristic $v_{OD}(v_{ID})$

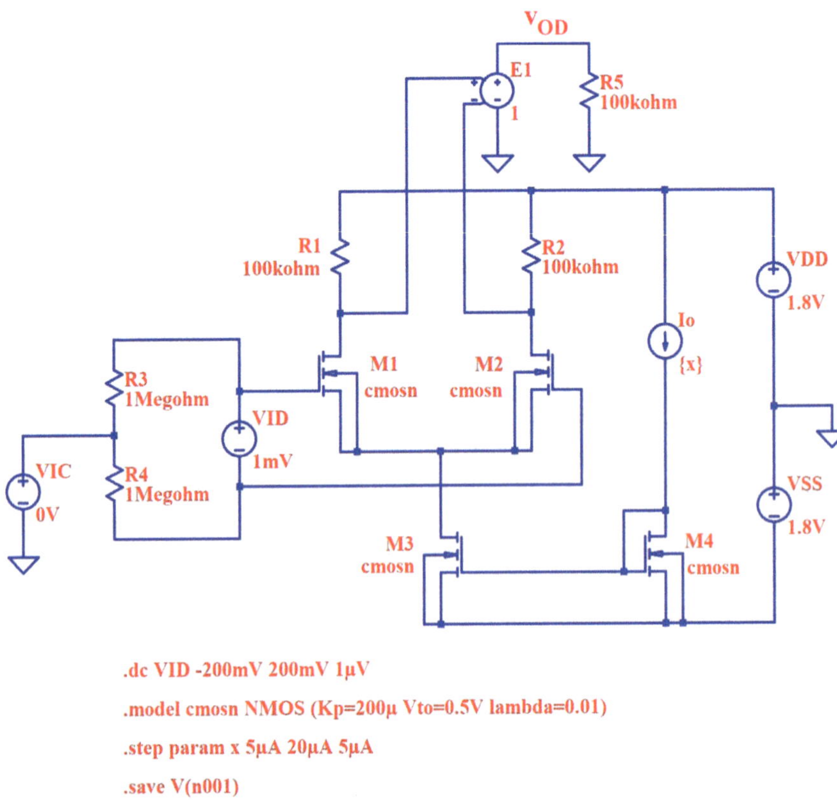

Fig. 2.14 The circuit used for simulating the transfer characteristics $v_{OD}(v_{ID})$, having as parameter the biasing current I_O

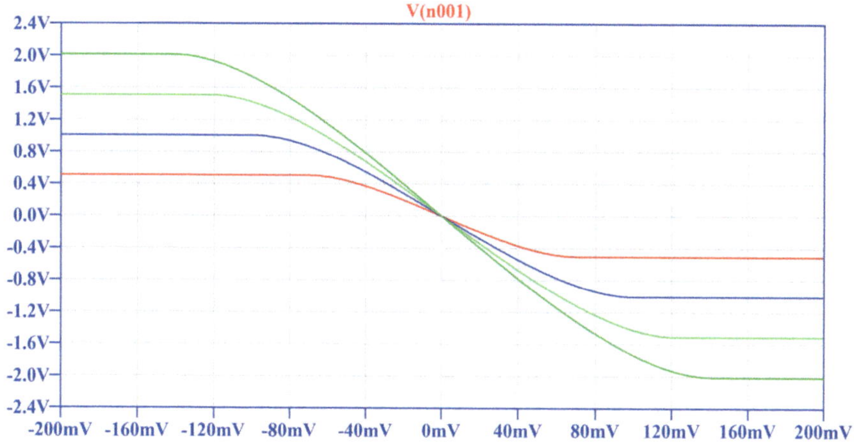

Fig. 2.15 The transfer characteristics $v_{OD}(v_{ID})$, having as parameter the biasing current I_O

2.2.5.2 Simulations for Evaluating the Differential-Mode Voltage Gain A_{DD}

In order to determine the differential-mode voltage gain A_{DD} of the basic NMOS passive load differential amplifier from Fig. 2.6, a *DC sweep analysis* is performed in LTspice, for a variation of the differential-mode input voltage v_{ID} between -1 mV and 1 mV. The small value of this voltage is set for allowing the differential amplifier to work in the quasi-linear region. The circuit used for this simulation is presented in Fig. 2.16, and the simulation results are shown in Fig. 2.17. In order to directly view the result of the simulation, the directive ".save V(n001)" is introduced in the file, with V(n001) representing the differential-mode output voltage of the circuit v_{OD}.

The additional circuit containing resistors R_3 and R_4 is used for generating the input voltages v_{I1} and v_{I2} of the differential amplifier, expressed by Eqs. (2.27) and (2.28) with $v_{IC} = 0$. Because resistors R_3 and R_4 are identical and the current flowing through them is the same, it results that the voltage across R_3 is equal to the voltage across R_4 and both are equal to $\frac{v_{ID}}{2}$. In consequence, the input voltages v_{I1} and v_{I2} can be expressed as follows:

$$v_{I1} = \frac{v_{ID}}{2} \tag{2.77}$$

and

$$v_{I2} = -\frac{v_{ID}}{2}. \tag{2.78}$$

The simulated value of the differential-mode voltage gain can be determined from Fig. 2.17:

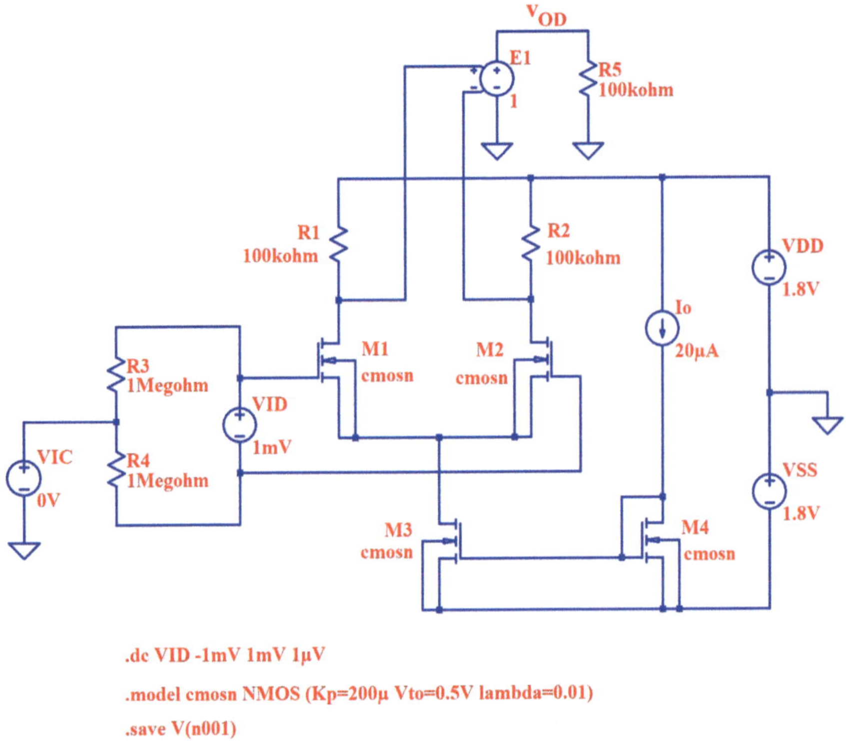

Fig. 2.16 The circuit used for simulating the small signal dependence $v_{OD}(v_{ID})$

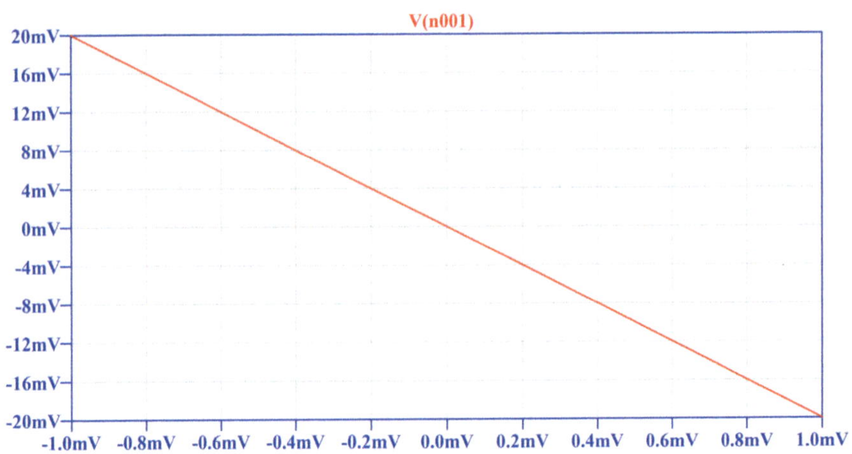

Fig. 2.17 Simulated small signal dependence $v_{OD}(v_{ID})$

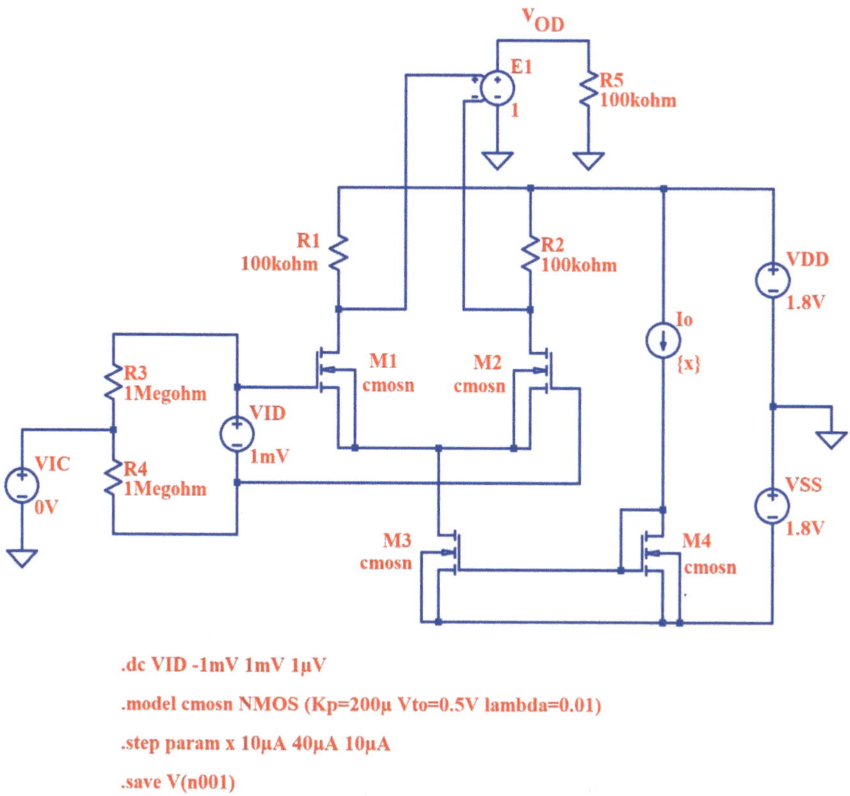

Fig. 2.18 The circuit used for simulating the small signal characteristics $v_{OD}(v_{ID})$, considering as parameter the biasing current I_O

$$A_{DD} = -\frac{(-20\,\text{mV}) - 20\,\text{mV}}{1\,\text{mV} - (-1\,\text{mV})} = -20, \tag{2.79}$$

identical to the theoretical value, computed using Eq. (2.73). For illustrating the influence of the biasing current I_O on the value of the differential-mode voltage gain A_{DD}, a *DC sweep analysis* is performed in LTspice, for a variation range of the voltage v_{ID} between -1 mV and 1 mV, and also a parametrical simulation for $x = I_O$, between 10 µA and 40 µA, with a step of 10 µA. The circuit used for this simulation is presented in Fig. 2.18, and the simulation results are shown in Fig. 2.19. In order to directly view the results of the simulation, the directive ".save V(n001)" is introduced in the file, with V(n001) representing the differential-mode output voltage of the circuit v_{OD}.

For illustrating the influence of the load resistances R_1 and R_2 on the value of the differential-mode voltage gain A_{DD}, a *DC sweep analysis* is performed in LTspice, for a variation range of the voltage v_{ID} between -1 mV and 1 mV, and also a parametrical simulation for $x = R_1 = R_2$, between 60 kΩ and 120 kΩ, with a step

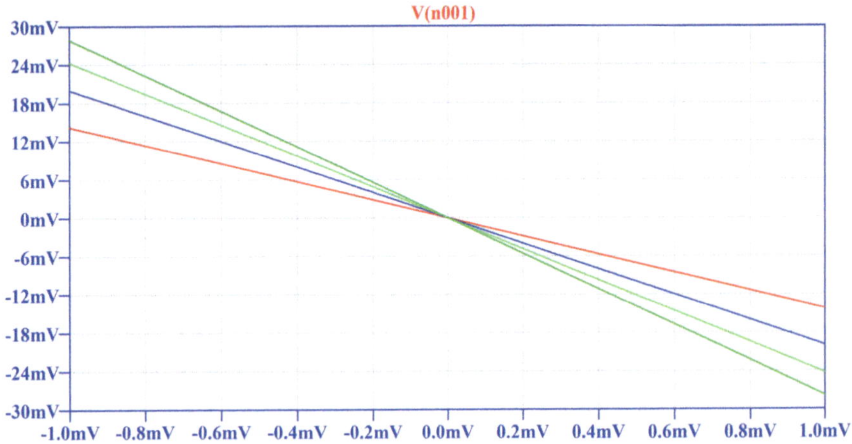

Fig. 2.19 Simulated small signal characteristics $v_{OD}(v_{ID})$, considering as parameter the biasing current I_O

of 20 kΩ. The circuit used for this simulation is presented in Fig. 2.20, and the simulation results are shown in Fig. 2.21. In order to directly view the result of the simulation, the directive ".save V(n001)" is introduced in the file, with V(n001) representing the differential-mode output voltage of the circuit v_{OD}.

2.2.5.3 Simulations for Evaluating the Maximal Range of the Common-Mode Input Voltage, $\left[V_{IC}^{\text{MIN}}, V_{IC}^{\text{MAX}}\right]$

In order to determine the simulated value of the maximal common-mode input voltage V_{IC}^{MAX}, a *DC sweep analysis* is performed in LTspice, for a variation of the voltage v_{ID} between -1 mV and 1 mV, and also a parametrical simulation for $x = V_{IC}$, with the parameter x having three values in the neighbourhood of the theoretical value $V_{IC}^{\text{MAX}} = 1.3$ V (Eq. (2.75)): 1.27 V, 1.28 V and 1.29 V. The circuit used for this simulation is presented in Fig. 2.22, and the simulation results are shown in Fig. 2.23. In order to directly view the results of the simulation, the directive ".save V(n001)" is introduced in the file, with V(n001) representing the differential-mode output voltage of the circuit v_{OD}.

The simulated value of V_{IC}^{MAX} is approximately 1.27 V, very close to the theoretical value expressed by Eq. (2.75).

For applications that require an operation of the NMOS differential amplifier with passive load at large values of the common-mode input voltage V_{IC}, close to its positive supply voltage $V_{DD} = 1.8$ V, two possible methods for increasing the value of the maximal common-mode input voltage V_{IC}^{MAX} can be used. Analysing the mathematical expression (2.49) of V_{IC}^{MAX}, these methods are as follows:

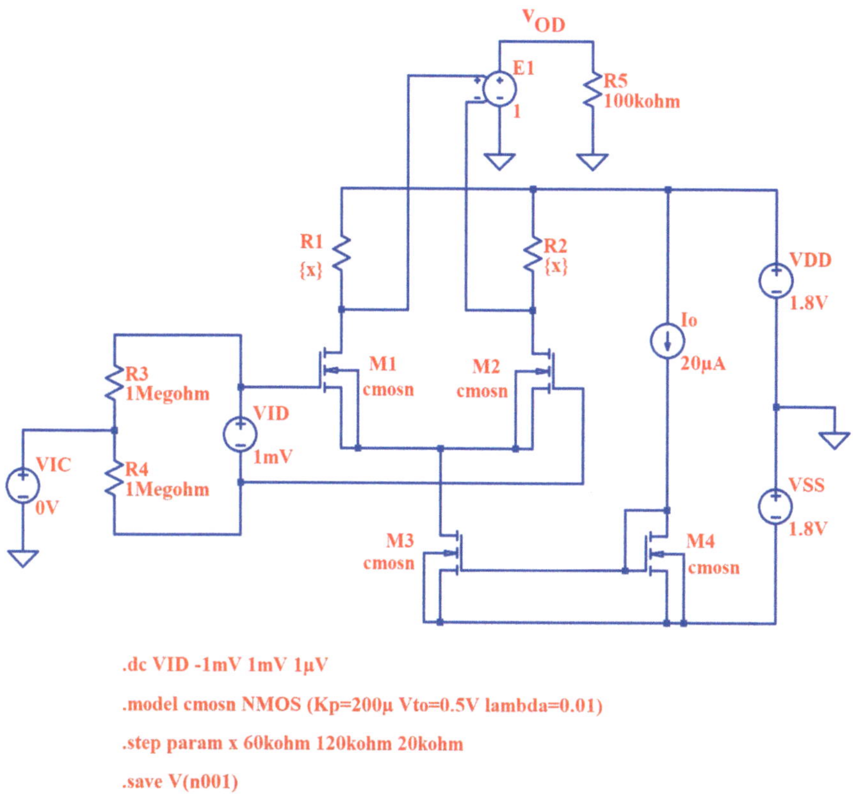

Fig. 2.20 The circuit used for simulating the small signal characteristics $v_{OD}(v_{ID})$, considering as parameter the load resistances $R_1 = R_2$

First Method: The Increasing of V_{IC}^{MAX} by Reducing the Load Resistance R_1

For illustrating this method, the resistance R_1 is reduced from the value of 100 kΩ to the value 50 kΩ, maintaining the biasing current to the initial value of 20 µA. The new value of the common-mode input voltage V_{IC}^{MAX} can be determined using Eq. (2.49):

$$V_{IC}^{\text{MAX}} = 1.8\,\text{V} - \frac{20\,\mu\text{A}}{2} \times 50\,\text{k}\Omega + 0.5\,\text{V} = 1.8\text{V}. \tag{2.80}$$

The cost of increasing the maximal common-mode input voltage V_{IC}^{MAX} from 1.3 V to 1.8 V using this design method is a reduction in the differential-mode voltage gain $A_{DD} = -g_{m1}(R_1//r_{ds1})$ (Eq. (2.43)), in the hypothesis of a differential-mode output voltage. Because the values of g_{m1} and r_{ds1} remain the same when changing the value of the load resistance R_1, the new value of the differential-mode voltage gain A_{DD} will be

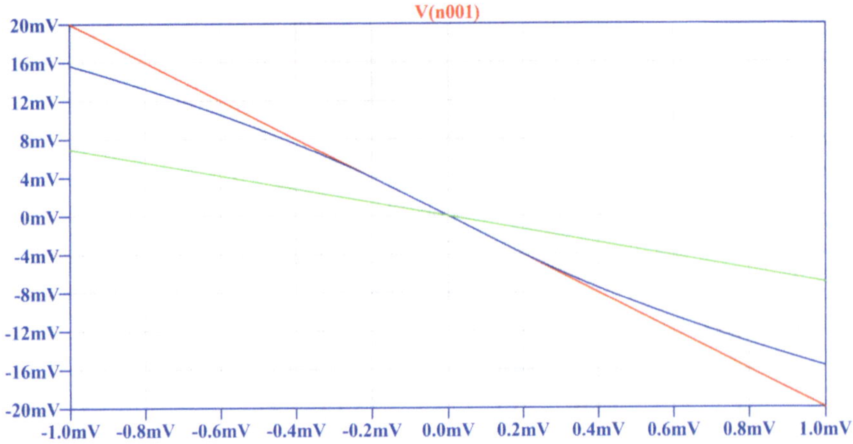

Fig. 2.23 Simulation results for determining the simulated value of the maximal common-mode input voltage V_{IC}^{MAX}—parametrical simulation $v_{OD}(v_{ID})$

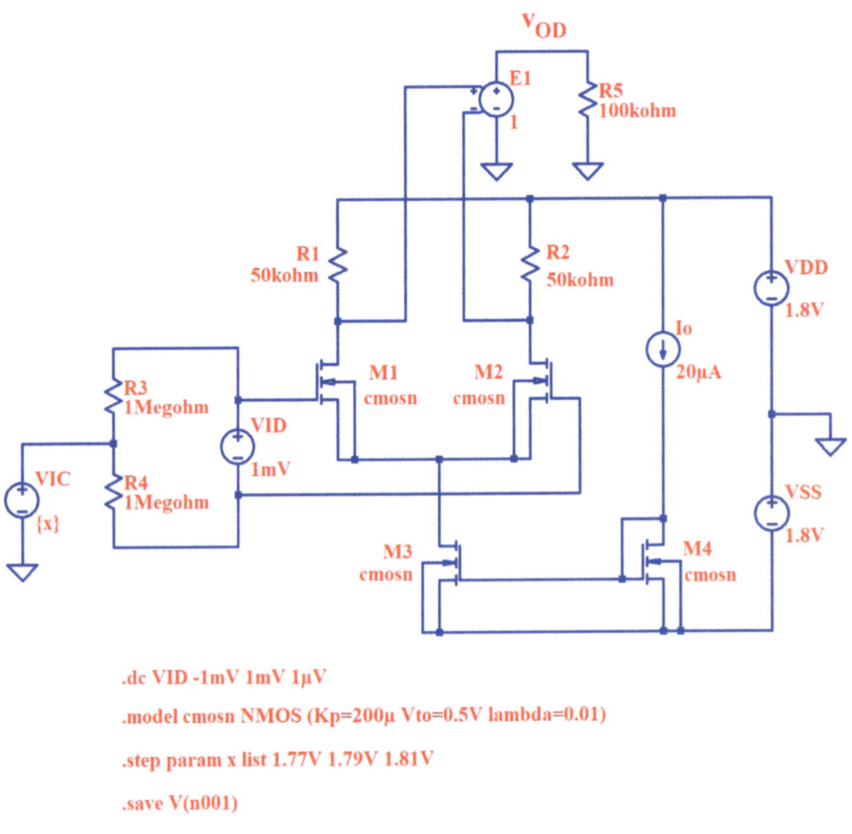

Fig. 2.24 Circuit for determining the simulated value of the increased maximal common-mode input voltage V_{IC}^{MAX} using the first method—parametrical simulation $v_{OD}(v_{ID})$

$$g_{m1} = \sqrt{2 \times 200\,\mu\text{A}/V^2 \times 10 \times 5\,\mu\text{A}} = 141.4\,\mu\text{A/V}, \tag{2.83}$$

$$r_{ds1} = \frac{1}{0.01\text{V}^{-1} \times 5\,\mu\text{A}} = 20\,\text{M}\Omega. \tag{2.84}$$

The new theoretical value of A_{DD} will be

$$A_{DD} = -141.4\,\mu\text{A/V} \times (100\,\text{k}\Omega//20\,\text{M}\Omega) \cong -14, \tag{2.85}$$

in comparison to with the previous value $A_{DD} = -20$, determined using Eq. (2.73).

In order to determine the simulated value of the new maximal common-mode input voltage V_{IC}^{MAX}, a *DC sweep analysis* is performed in LTspice, for a variation of the voltage v_{ID} between -1 mV and 1 mV, and also a parametrical simulation for $x = V_{IC}$, with the parameter x having three values in the neighbourhood of the new theoretical value $V_{IC}^{\text{MAX}} = 1.8\,\text{V}$: 1.77 V, 1.79 V and 1.81 V. The circuit used for this simulation is presented in Fig. 2.26, and the simulation results are shown in Fig. 2.27. In order to directly view the results of the simulation, the directive ". save V(n001)" is introduced in the file, with V(n001) representing the differential-mode output voltage of the circuit v_{OD}.

The simulated value of the maximal common-mode input voltage V_{IC} is approximately 1.77 V, very close to the theoretical value, determined using Eq. (2.82).

The maximal absolute value of the slope of the characteristics $v_{OD}(v_{ID})$ from Fig. 2.27 corresponds to a common-mode input voltage still in the maximal range $\left[V_{IC}^{\text{MIN}}, V_{IC}^{\text{MAX}}\right]$. The maximal absolute value of 14 of this slope is equal to the new reduced absolute value of the differential-mode voltage gain A_{DD}, determined using Eq. (2.85). The other two characteristics from Fig. 2.27, having absolute values of their slopes inferior to the value of 10, correspond to a common-mode input voltage V_{IC} outside the interval $\left[V_{IC}^{\text{MIN}}, V_{IC}^{\text{MAX}}\right]$ ($V_{IC} = 1.79\,\text{V}$ and $V_{IC} = 1.81\,\text{V}$, respectively), equivalent to the transistors M1 and M2 no longer biased in saturation region, because $V_{IC} > V_{IC}^{\text{MAX}}$ for these values of V_{IC}.

In conclusion, in order to increase the maximal value of the common-mode input voltage V_{IC}^{MAX}, the second method is better than the first method. Its cost related to the reduction in the differential-mode voltage gain A_{DD} is smaller (from -20 to -14), compared with the reduction from -20 to -10 for the first method, with both methods having the same benefit of increasing V_{IC}^{MAX} from 1.3 V to 1.8 V.

In order to determine the simulated value of the minimal common-mode input voltage V_{IC}^{MIN}, a *DC sweep analysis* is performed in LTspice, for a variation of the voltage v_{ID} between -1 mV and 1 mV, and also a parametrical simulation for $x = V_{IC}$, with the parameter x having three values in the neighbourhood of the theoretical value $V_{IC}^{\text{MIN}} = -1.059\,\text{V}$ (Eq. (2.76)): -1.06 V, -1.1 V and -1.14 V. The circuit used for this simulation is presented in Fig. 2.28, and the simulation results are shown in Fig. 2.29. In order to directly view the results of the simulation, the directive ".save V(n001)" is introduced in the file, with V(n001) representing the differential-mode output voltage of the circuit v_{OD}.

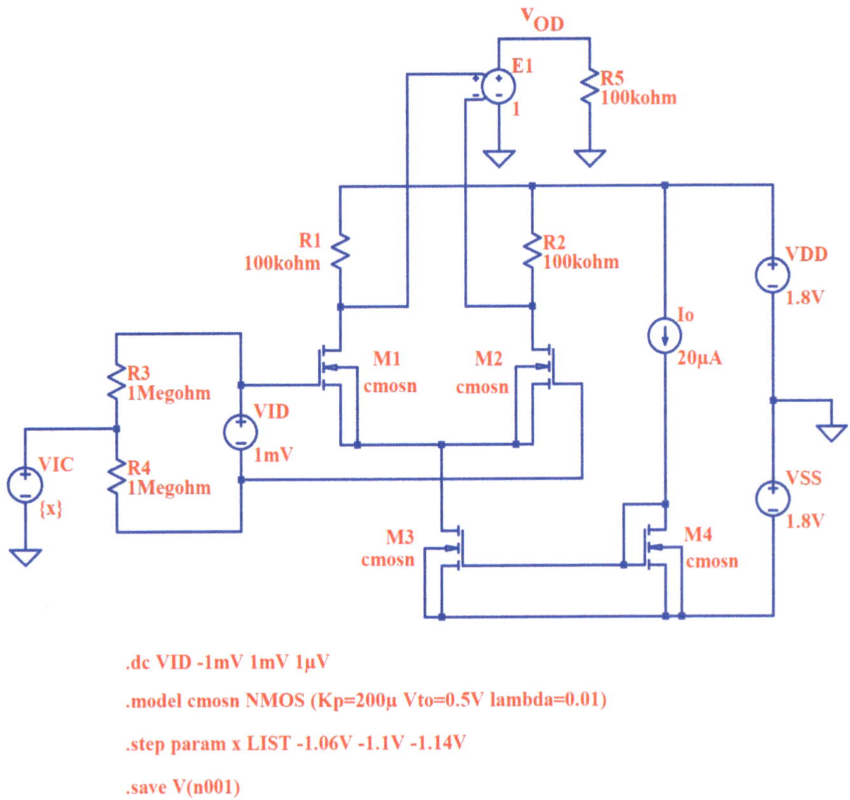

Fig. 2.28 Circuit for determining the simulated value of the minimal common-mode input voltage V_{IC}^{MIN}—parametrical simulation $v_{OD}(v_{ID})$

a step of 0.01 V^{-1} is added to the previous *DC sweep analysis*, for a variation of the common-mode input voltage V_{IC} between -1 V and 1 V. The circuit used for this simulation is presented in Fig. 2.32, and the simulation results are shown in Fig. 2.33. In order to directly view the result of the simulation, the directive ".save V(n002)" is introduced in the file, with V(n002) representing the common-mode output voltage of the circuit v_{OC}.

The absolute value of the common-mode voltage gain A_{CC} (which is equal to the absolute value of the slope of the characteristics shown in Fig. 2.33) increases when the channel-length modulation parameter λ_n increases, from 0.01 V^{-1} to 0.05 V^{-1}.

For decreasing the absolute value of the common-mode voltage gain A_{CC}, the current mirror M3–M4 that biases the differential amplifier presented in Fig. 2.30 is replaced in Fig. 2.34 with a cascode current mirror M3–M6. The utilization of the cascode current mirror will strongly increase the output resistance of the current source, producing a very important decreasing of the absolute value of the common-mode voltage gain A_{CC}.

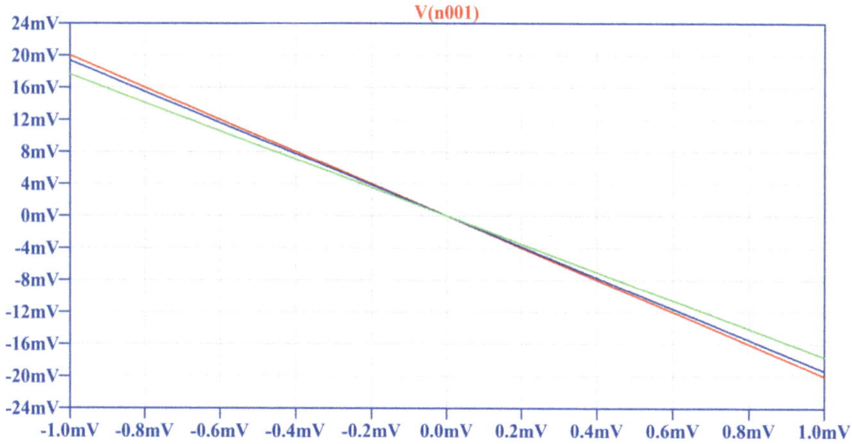

Fig. 2.29 Simulation results for determining the simulated value of the minimal common-mode input voltage V_{IC}^{MIN}—parametrical simulation $v_{OD}(v_{ID})$

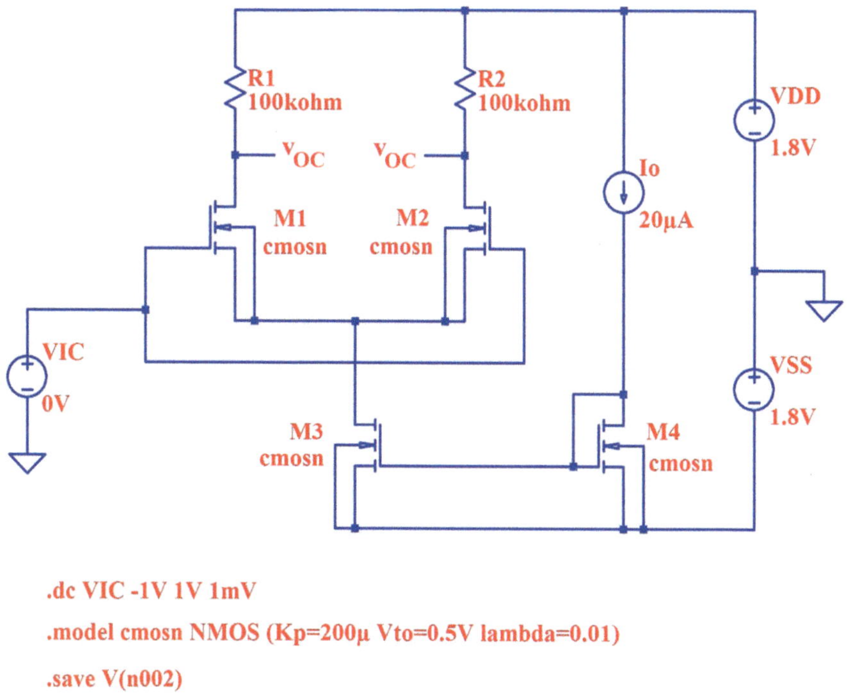

Fig. 2.30 The circuit used for simulating the dependence $v_{OC}(v_{IC})$ (for the NMOS differential amplifier biased using a basic current mirror)

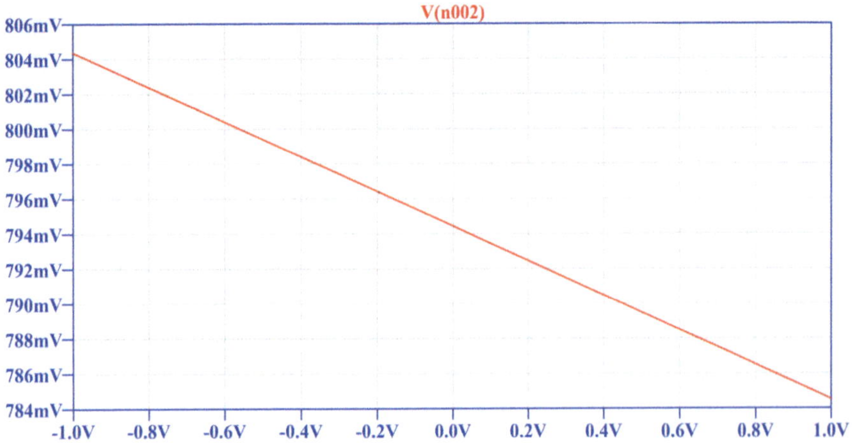

Fig. 2.31 Simulated characteristic $v_{OC}(v_{IC})$ (for the NMOS differential amplifier biased using a basic current mirror)

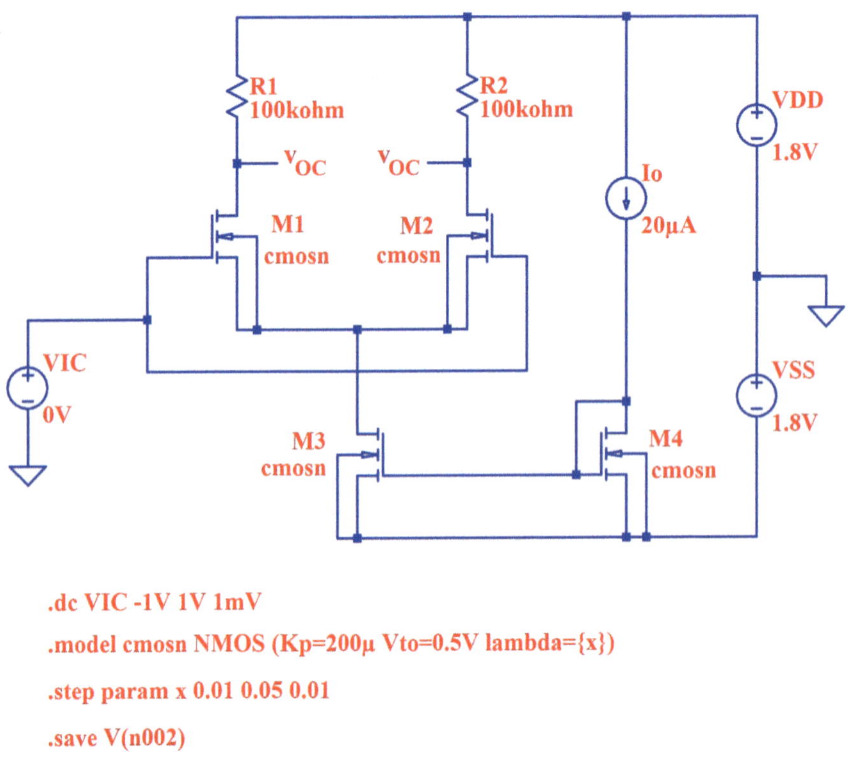

Fig. 2.32 Circuit used for simulating the dependence $v_{OC}(v_{IC})$, with an additional parametrical simulation for $x = \lambda_n$ (for the NMOS differential amplifier biased using a basic current mirror)

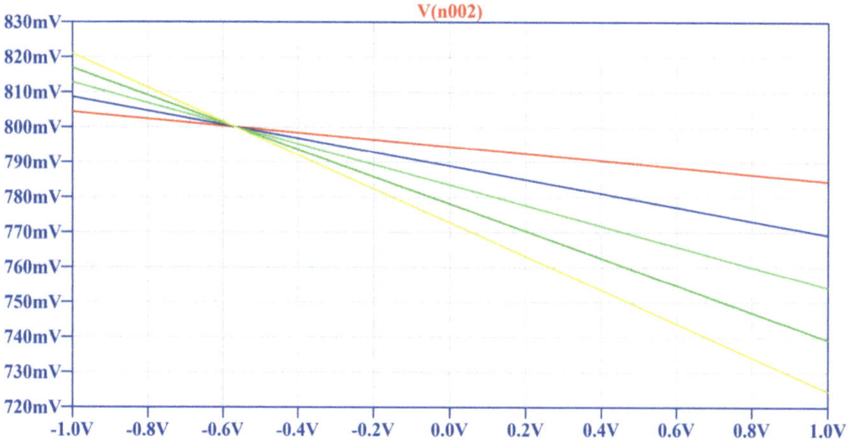

Fig. 2.33 Simulated characteristic $v_{OC}(v_{IC})$, with an additional parametrical simulation for $x = \lambda_n$ (for the NMOS differential amplifier biased using a basic current mirror)

In order to determine the common-mode voltage gain A_{CC} of the basic NMOS passive load differential amplifier, biased using a cascode current mirror from Fig. 2.34, a *DC sweep analysis* is performed in LTspice, for a variation of the common-mode input voltage v_{IC} between $-0.4\,\text{V}$ and $1\,\text{V}$. This interval $[-0.4\,\text{V}, 1\,\text{V}]$ is set for a proper operation of the differential amplifier biased at a cascode current mirror, so it must be included in the new more restrictive maximal interval defined by $\left[V_{IC}^{\text{MIN}}, V_{IC}^{\text{MAX}}\right]$. It can be observed that only the minimal value of the common-mode input voltage V_{IC}^{MIN} has been changed when replacing the classical current mirror from Fig. 2.30 with the cascode current mirror from Fig. 2.34.

The circuit used for this simulation is presented in Fig. 2.34, and the simulation results are shown in Fig. 2.35. In order to directly view the result of the simulation, the directive ".save V(n002)" is introduced in the file, with V(n002) representing the common-mode output voltage of the circuit v_{OC}.

The simulated value of the common-mode voltage gain can be determined from Fig. 2.35:

$$A_{CC} = \frac{-10\,\mu\text{V}}{1.4\,\text{V}} = -7.14 \times 10^{-6}. \tag{2.87}$$

Similarly to the method used for determining the theoretical expression of A_{CC} for the differential amplifier biased using a basic current mirror presented in Fig. 2.30, the expression of the common-mode voltage gain for the differential amplifier biased at a cascode current mirror from Fig. 2.34 can be expressed as follows:

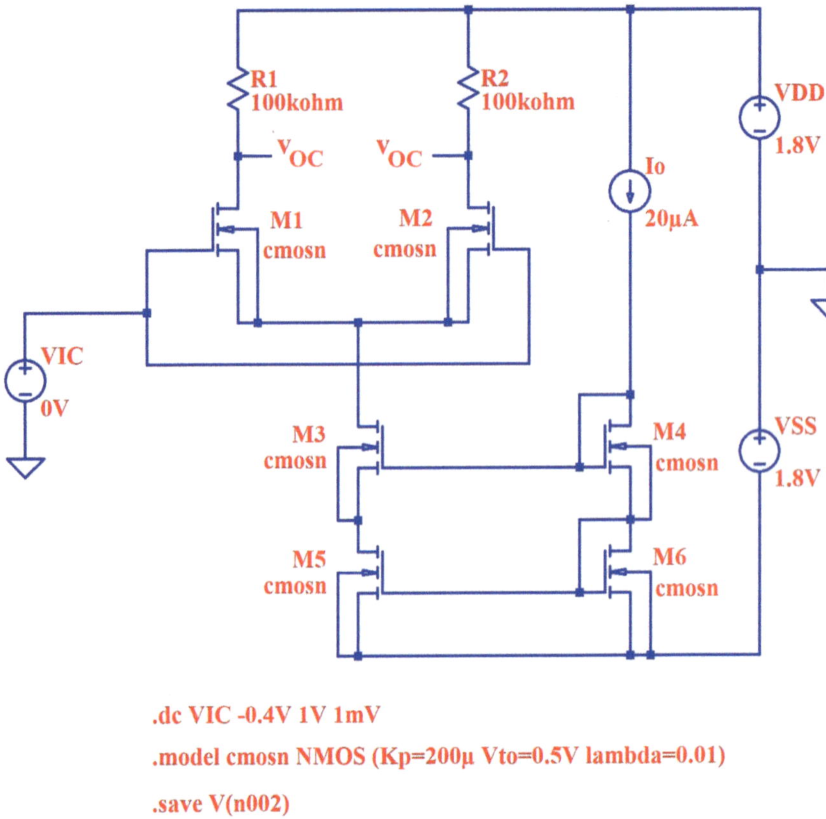

Fig. 2.34 The circuit used for simulating the dependence $v_{OC}(v_{IC})$ when biasing the NMOS differential amplifier using a cascode current mirror

$$A_{CC} \cong -\frac{R_1}{2g_{m3}r_{ds3}^2}. \tag{2.88}$$

The value of transconductance g_{m3} is

$$g_{m3} = \sqrt{2 \times 200\,\mu\text{A/V}^2 \times 10 \times 20\,\mu\text{A}} = 282.84\,\mu\text{A/V} \tag{2.89}$$

and the value of r_{ds3} is

$$r_{ds3} = \frac{1}{0.01\,\text{V}^{-1} \times 20\,\mu\text{A}} = 5\,\text{M}\Omega, \tag{2.90}$$

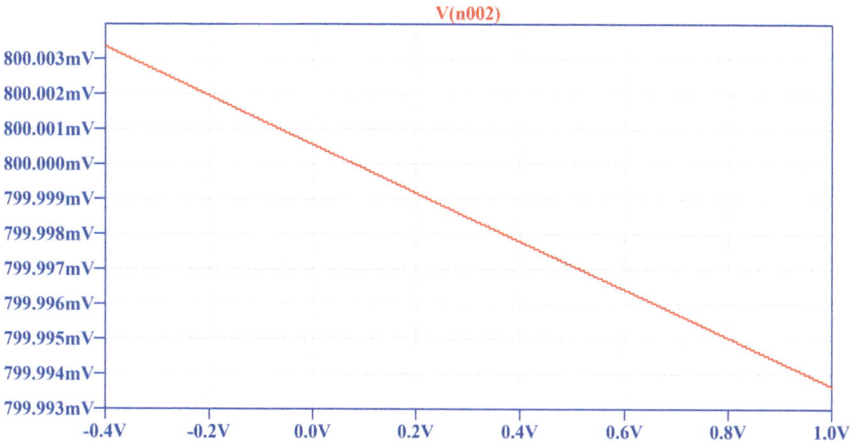

Fig. 2.35 Simulated characteristic $v_{OC}(v_{IC})$ when biasing the NMOS differential amplifier using a cascode current mirror

resulting in the following theoretical value of A_{CC}:

$$A_{CC} = -\frac{100\,\text{k}\Omega}{2 \times 282.84\,\mu\text{A/V} \times 25(\text{M}\Omega)^2} = -7.07 \times 10^{-6}, \tag{2.91}$$

very close to the simulated value of -7.14×10^{-6}, obtained using Eq. (2.87).

In order to evaluate the dependence of the common-mode voltage gain for the differential amplifier biased at a cascode current mirror on the technological parameter λ_n, which models the channel-length modulation effect, an additional parametrical simulation for $x = \lambda_n$, between $0.01\,\text{V}^{-1}$ and $0.05\,\text{V}^{-1}$, with a step of $0.01\,\text{V}^{-1}$ is added to the previous *DC sweep analysis*, for a variation of the common-mode input voltage V_{IC} between -0.4 V and 1 V. The circuit used for this simulation is presented in Fig. 2.36, and the simulation results are shown in Fig. 2.37. In order to directly view the result of the simulation, the directive ".save V(n002)" is introduced in the file, with V(n002) representing the common-mode output voltage of the circuit v_{OC}.

The absolute value of the common-mode voltage gain A_{CC} (which is equal to the absolute value of the slope of the characteristics shown in Fig. 2.37) increases when the channel-length modulation parameter λ_n increases, from $0.01\,\text{V}^{-1}$ to $0.05\,\text{V}^{-1}$.

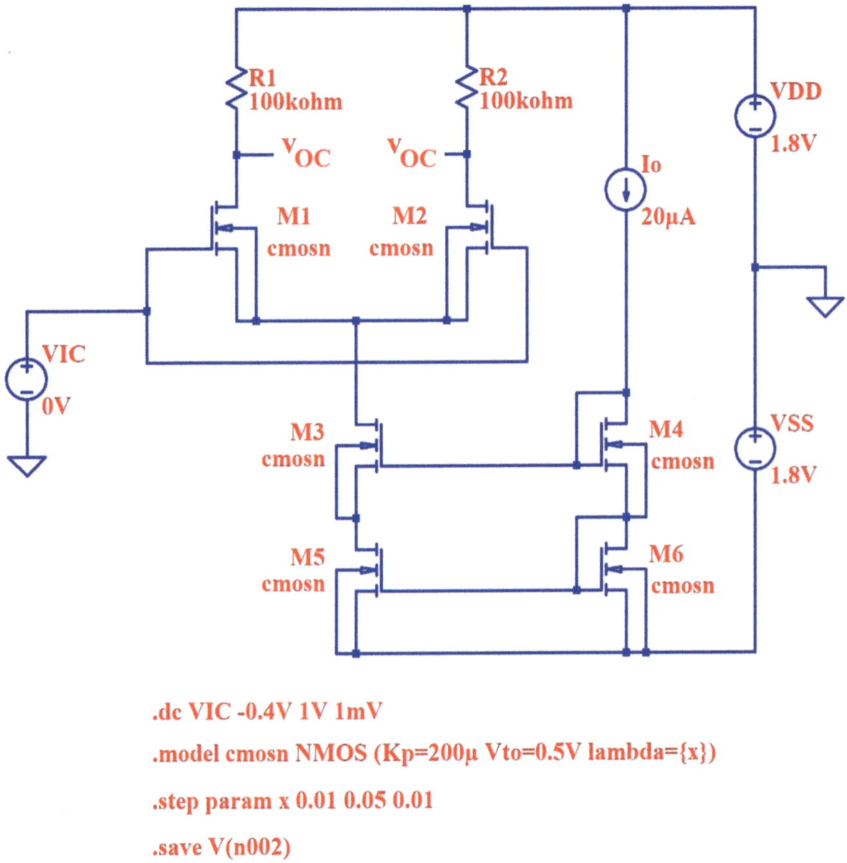

Fig. 2.36 Circuit used for simulating the dependence $v_{OC}(v_{IC})$, with an additional parametrical simulation for $x = \lambda_n$ (for the NMOS differential amplifier biased at a cascode current mirror)

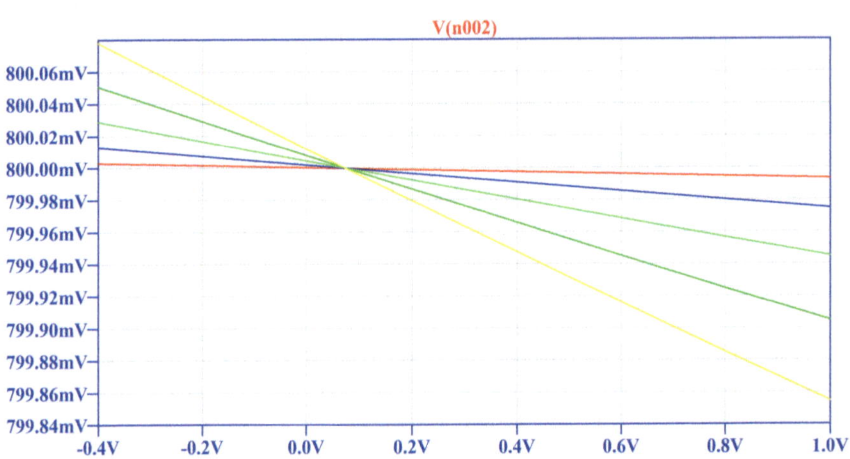

Fig. 2.37 Simulated characteristic $v_{OC}(v_{IC})$, with an additional parametrical simulation for $x = \lambda_n$ (for the NMOS differential amplifier biased at a cascode current mirror)

2.3 Basic PMOS Differential Amplifier with Passive Load

The basic PMOS differential amplifier with passive load, having a theoretical perfect symmetrical structure, represents the complementary circuit with respect to the basic NMOS differential amplifier with passive load. It must ideally amplify only the differential-mode input voltage and must reject the common-mode input voltage.

The circuit of the PMOS differential amplifier is presented in Fig. 2.38. It consists of two identical PMOS transistors (M1 and M2) and two identical load resistances (R_1 and R_2). The PMOS differential amplifier with passive load is biased using a current source I_O, having an output resistance R_O.

2.3.1 *Large Signal Analysis*

The large signal analysis of the PMOS differential amplifier is similar with the same analysis (presented in Sect. 2.2.1), made for the complementary basic NMOS differential amplifier, shown in Fig. 2.1. The conclusion is the same: the transfer characteristic $v_{OD}(v_{ID})$ of the PMOS differential amplifier is strongly non-linear, so the circuit presented in Fig. 2.38 can be used for amplifying signals only for a restricted range of the differential-mode input voltage v_{ID}.

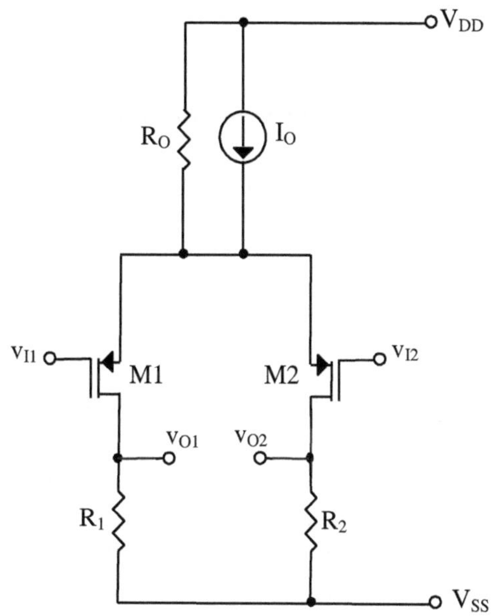

Fig. 2.38 Basic PMOS differential amplifier with passive load

2.3.2 *Small Signal Analysis*

Also the small signal analysis of the PMOS differential amplifier with passive load is similar with the same analysis (presented in Sect. 2.2.2), made for the complementary basic NMOS differential amplifier with passive load, shown in Fig. 2.1.

2.3.2.1 Computation of the Differential-Mode Voltage Gain A_{DD} Using the Half-Circuit Method

Using a similar half-circuit method used for determining the differential-mode voltage gain of the PMOS differential amplifier with passive load, the conclusion will be the following:

- For the utilization of the voltage $v_{O1} - v_{O2} = v_{OD}$ as the output voltage of the differential amplifier (differential-mode output voltage), the differential-mode voltage gain of the basic PMOS differential amplifier with passive load is equal to the voltage gain of the differential-mode half circuit, $A_{DD} = A_{DD\,-\,HC} = \; - g_{m1}(R_1 // r_{ds1})$.
- For the utilization of the voltage $v_{O1} = \frac{v_{OD}}{2}$ or voltage $v_{O2} = -\frac{v_{OD}}{2}$ as the output voltage of the differential amplifier (single-mode output voltage), the differential-mode voltage gain of the basic PMOS differential amplifier with passive load is equal to half of the voltage gain of the differential-mode half circuit, $A_{DD} = \frac{A_{DD-HC}}{2} = -\frac{1}{2} g_{m1}(R_1 // r_{ds1})$.

2.3.2.2 Computation of the Common-Mode Voltage Gain A_{CC} Using the Half-Circuit Method

Using a similar half-circuit method, the common-mode voltage gain of the PMOS differential amplifier with passive load will be

$$A_{CC} = \frac{v_{OC}}{v_{IC}} = -\frac{g_{m1}R_1}{1 + 2g_{m1}R_O} \cong -\frac{R_1}{2R_O}. \tag{2.92}$$

2.3.3 *Maximal Range of the Common-Mode Input Voltage*

A differential amplifier must amplify as much as possible the differential-mode input voltage v_{ID} and must reject as much as possible the common-mode input voltage v_{IC}. Another problem related to the common-mode input voltage is the maximal range of this voltage $\left[V_{IC}^{\mathrm{MIN}}, V_{IC}^{\mathrm{MAX}}\right]$. If the common-mode voltage applied on the inputs of a differential amplifier is superior to the maximal acceptable value V_{IC}^{MAX}, or inferior to the minimal acceptable value V_{IC}^{MIN}, a part of the circuit does no longer work and all the differential amplifier operation is compromised. For this reason, it is extremely important to determine the theoretical limits of this maximal range of the common-

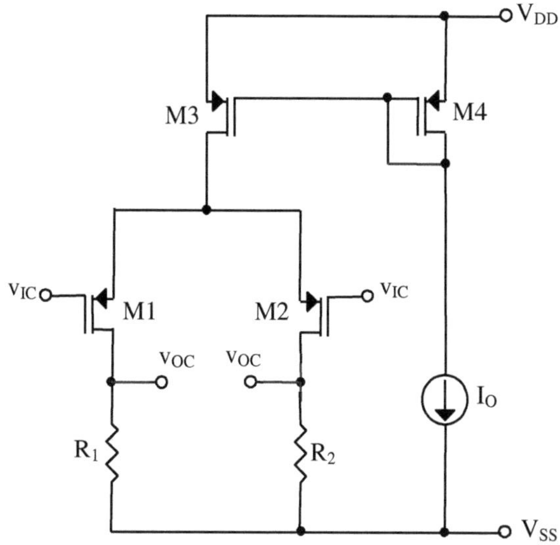

Fig. 2.39 Basic PMOS differential amplifier with passive load, biased using a basic MOS current mirror

mode input voltage and to use the differential amplifier only for common-mode input voltages that are inside this interval.

In order to determine the maximal range of the common-mode input voltage for the basic PMOS differential amplifier with passive load, it is necessary to choose a particular implementation of the current source I_O for biasing the differential amplifier presented in Fig. 2.38. Because the most used current sources from Chap. 1 are represented by the MOS basic current mirror from Fig. 1.16, a complementary version of this current source is used in Fig. 2.39 for biasing the basic PMOS differential amplifier with passive load.

The maximal value of the common-mode input voltage is limited by the operation in saturation of transistor M3:

$$V_{IC}^{MAX} = V_{DD} - (V_{SG3} - V_T) - V_{SG1}. \tag{2.93}$$

Using the approximate squaring characteristic of transistors M1 and M3, biased in saturation region, the previous relation becomes

$$V_{IC}^{MAX} = V_{DD} - V_T - \left(\sqrt{2} + 1\right)\sqrt{\frac{I_O}{K'_p\left(\frac{W}{L}\right)}}. \tag{2.94}$$

The only method for improving the behavior of the basic PMOS differential amplifier with passive load from the perspective of its maximal common-mode input voltage is to decrease the biasing current I_O. The disadvantage of this approach is the reducing of the differential-mode voltage gain A_{DD}, because of the reduction of transconductances g_{m1} and g_{m2} of transistors M1 and M2. Even accepting these disadvantages, the value of V_{IC}^{MAX} is still strongly limited in comparison to the value of V_{IC}^{MIN}.

The minimal value of the common-mode input voltage is limited by the operation in saturation of both transistors M1 and M2:

$$V_{IC}^{\mathrm{MIN}} = -V_{SS} + \frac{I_O}{2} R_1 + (V_{SG1} - V_T) - V_{SG1} = -V_{SS} + \frac{I_O}{2} R_1 - V_T. \quad (2.95)$$

The decreasing of V_{IC}^{MIN} is possible by decreasing the load resistance R_1 or by decreasing the biasing current I_O. Unfortunately, both methods will decrease the differential-mode voltage gain A_{DD}. Sacrificing this gain, it is theoretically possible to obtain $V_{IC}^{\mathrm{MIN}} < -V_{SS}$, having the significance that the differential amplifier is extremely tolerant: it can accept a common-mode input voltage inferior to the value of its negative supply voltage (of course, this voltage can be applied at their inputs only by another circuit, which is supplied at a voltage inferior to $-V_{SS}$).

In conclusion, the basic PMOS differential amplifier with passive load can usually be used especially for relatively small values of the common-mode input voltages. Applications that need a proper operation of the differential amplifier for large values of the common-mode input voltage must replace the basic PMOS differential amplifier with passive load with its complementary circuit, the basic NMOS differential amplifier with passive load.

2.3.4 Input Offset Voltage

The significance and the method for determining the approximate expression of the input offset voltage for the basic PMOS differential amplifier with passive load are similar to the method presented for the basic NMOS differential amplifier with passive load (Eq. (2.69)).

2.3.5 Design Examples

In order to confirm the previously presented theoretical estimated results, the basic PMOS passive load differential amplifier from Fig. 2.39 is simulated with LTspice using the following values: $K'_p = 100\,\mu\mathrm{A/V}^2$, $\lambda_p = 0.02\,\mathrm{V}^{-1}$, $V_{Tp} = -0.5\,\mathrm{V}$, $V_{DD} = 1.8\,\mathrm{V}$ and $V_{SS} = -1.8\,\mathrm{V}$. MOS transistors have the aspect ratio $\left(\frac{W}{L}\right)_1 = \left(\frac{W}{L}\right)_2 = \left(\frac{W}{L}\right)_3 = \left(\frac{W}{L}\right)_4 = \frac{5\,\mu\mathrm{m}}{0.5\,\mu\mathrm{m}} = 10$. The values of resistances are $R_1 = R_2 = 100\,\mathrm{k}\Omega$, while the value of the current source is $I_O = 20\,\mu\mathrm{A}$.

Equation (2.13) can be used for determining the values of the differential-mode input voltage for complete current switching in one branch of the differential amplifier (left or right):

$$(v_{ID})_{1,2} = \pm \sqrt{\frac{2 \times 20\,\mu\mathrm{A}}{100\,\mu\mathrm{A/V}^2 \times 10}} \cong \pm 0.2\,\mathrm{V}. \quad (2.96)$$

In order to compute the theoretical value of the differential-mode voltage gain expressed using Eqs. (2.41) and (2.43), it is necessary to compute the transconductance g_{m1} and the internal resistance r_{ds1} of transistor M1:

$$g_{m1} = \sqrt{2 \times 100\,\mu\text{A/V}^2 \times 10 \times 10\,\mu\text{A}} = 141.4\,\mu\text{A/V}, \tag{2.97}$$

$$r_{ds1} = \frac{1}{0.02\,\text{V}^{-1} \times 10\,\mu\text{A}} = 5\,\text{M}\Omega. \tag{2.98}$$

The theoretical value of A_{DD} is

$$A_{DD} = -141.4\,\mu\text{A/V} \times (100\,\text{k}\Omega//5\,\text{M}\Omega) \cong -14.1. \tag{2.99}$$

The theoretical value of the common-mode voltage gain A_{CC} can be computed using Eq. (2.46) and replacing the output resistance R_O of the current mirror M3–M4 with r_{ds3}:

$$A_{CC} = -\frac{100\,\text{k}\Omega}{2 \times \frac{1}{0.02\,\text{V}^{-1} \times 20\,\mu\text{A}}} = -\frac{100\,\text{k}\Omega}{5\,\text{M}\Omega} = -0.02. \tag{2.100}$$

The theoretical value of the maximal value of the common-mode input voltage is expressed by Eq. (2.94):

$$V_{IC}^{MAX} = 1.8\,\text{V} - 0.5\,\text{V} - 2.41 \times \sqrt{\frac{20\,\mu\text{A}}{100\,\mu\text{A/V}^2 \times 10}} \cong 0.96\,\text{V}. \tag{2.101}$$

The theoretical value of the minimal common-mode input voltage is expressed by Eq. (2.95):

$$V_{IC}^{MIN} = -1.8\,\text{V} + \frac{20\,\mu\text{A}}{2} \times 100\,\text{k}\Omega - 0.5\,\text{V} = -1.3\,\text{V}. \tag{2.102}$$

2.3.5.1 Large Signal Operation Simulations

In order to graphically represent the dependencies of the drain currents of transistors M1 and M2 on the differential-mode input voltage v_{ID}, for the basic PMOS passive load differential amplifier from Fig. 2.39, a *DC sweep analysis* is performed in LTspice, for a variation of the voltage v_{ID} between −300 mV and 300 mV. The circuit used for this simulation is presented in Fig. 2.40 (it is necessary to set two current markers in LTspice, in the drains of transistors M1 and M2), and the simulation results are shown in Fig. 2.41. The source E1 is used for computing the differential-mode output voltage v_{OD}.

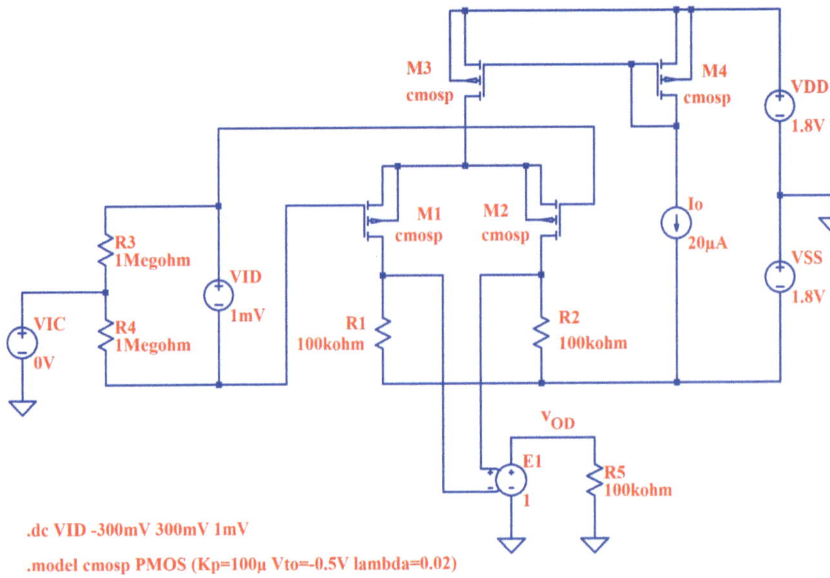

Fig. 2.40 The circuit used for simulating the dependencies $i_{D1}(v_{ID})$ and $i_{D2}(v_{ID})$

The simulated value of the differential-mode input voltage for complete current switching in one branch of the differential amplifier (left or right) is approximately equal to 0.2 V, identical to the theoretical value obtained in Eq. (2.96).

For illustrating the influence of the biasing current I_O on the operation of the differential amplifier presented in Fig. 2.39, a *DC sweep analysis* is performed in LTspice, for a variation range of the voltage v_{ID} between −300 mV and 300 mV, and also a parametrical simulation for $x = I_O$, between 5 μA and 20 μA, with a step of 5 μA. The circuit used for this simulation is presented in Fig. 2.42 (it is necessary to set two current markers in LTspice, in the drains of transistors M1 and M2), and the simulation results are shown in Fig. 2.43.

In order to graphically represent the transfer characteristic $v_{OD}(v_{ID})$ of the basic PMOS passive load differential amplifier from Fig. 2.39, a *DC sweep analysis* is performed in LTspice, for a variation range of the voltage v_{ID} between −300 mV and 300 mV. The circuit used for this simulation is presented in Fig. 2.44. The simulation results are shown in Fig. 2.45. In order to directly view the result of the simulation, the directive ".save V(n010)" is introduced in the file, with V(n010) representing the differential-mode output voltage of the circuit v_{OD}.

For illustrating the influence of the biasing current I_O on the operation of the differential amplifier presented in Fig. 2.39, a *DC sweep analysis* is performed in LTspice, for a variation range of the voltage v_{ID} between −300 mV and 300 mV, and also a parametrical simulation for $x = I_O$, between 5 μA and 20 μA, with a step of 5 μA. The circuit used for this simulation is presented in Fig. 2.46, and the simulation

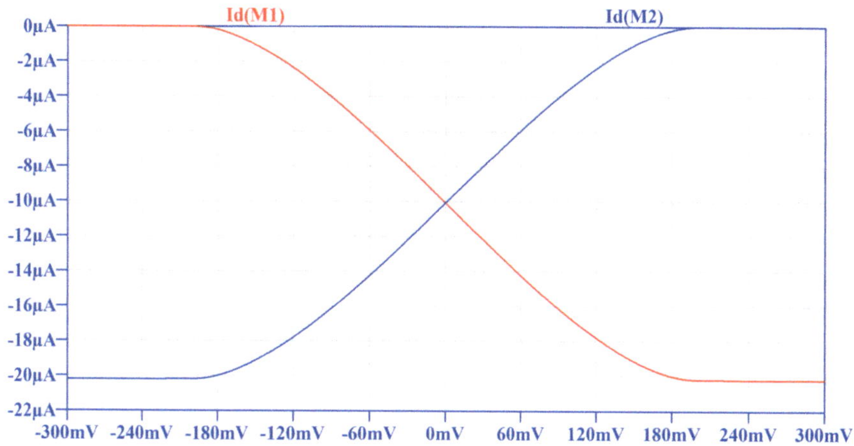

Fig. 2.41 The dependencies $i_{D1}(v_{ID})$ and $i_{D2}(v_{ID})$

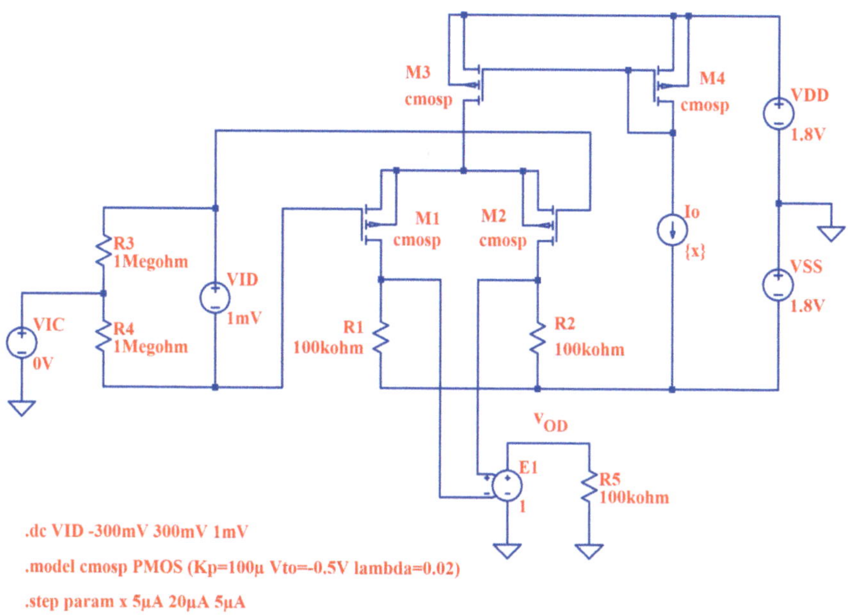

Fig. 2.42 The circuit used for simulating the dependencies $i_{D1}(v_{ID})$ and $i_{D2}(v_{ID})$, having as parameter the biasing current I_O

results are shown in Fig. 2.47. In order to directly view the results of the simulation, the directive ".save V(n010)" is introduced in the file, with V(n010) representing the differential-mode output voltage of the circuit v_{OD}.

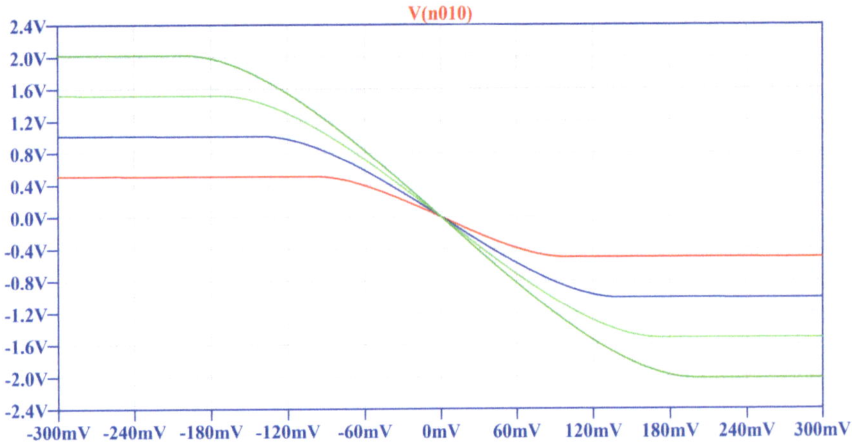

Fig. 2.47 The transfer characteristics $v_{OD}(v_{ID})$, having as parameter the biasing current I_O

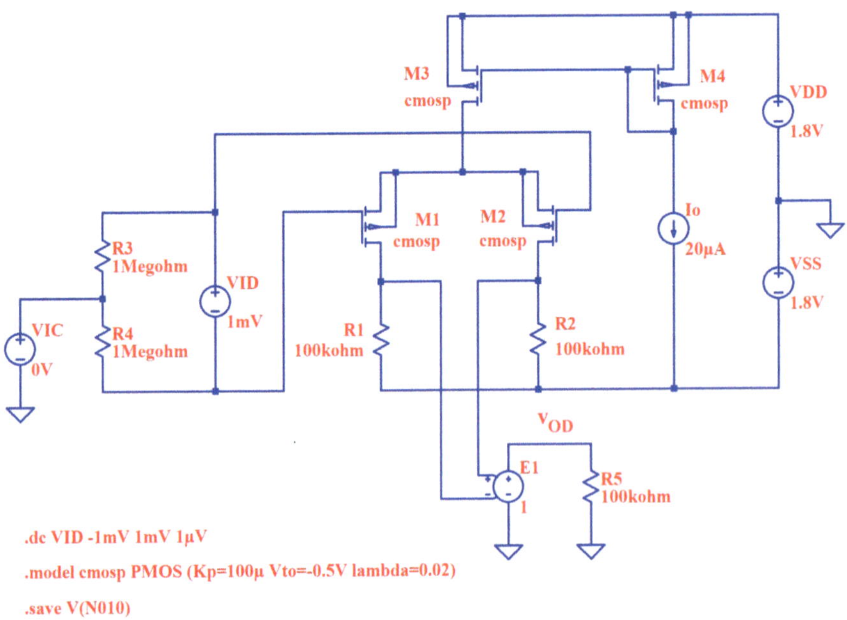

Fig. 2.48 The circuit used for simulating the small signal dependence $v_{OD}(v_{ID})$

the result of the simulation, the directive ".save V(n010)" is introduced in the file, with V(n010) representing the differential-mode output voltage of the circuit v_{OD}.

The simulated value of the differential-mode voltage gain can be determined from Fig. 2.49:

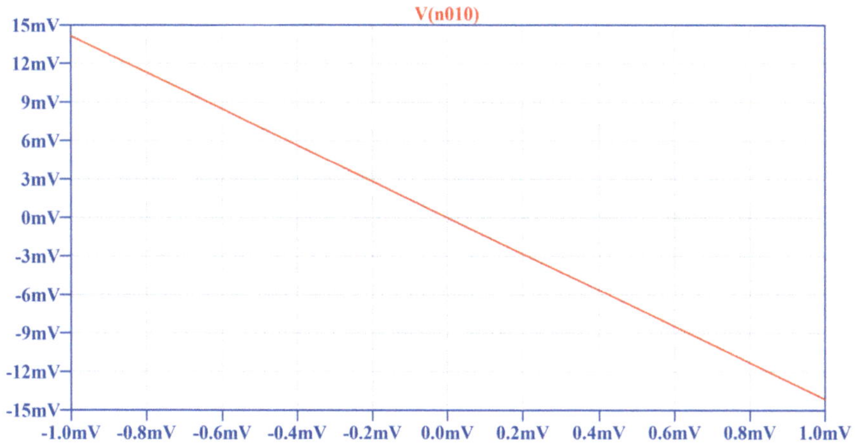

Fig. 2.49 Simulated small signal dependence $v_{OD}(v_{ID})$

$$A_{DD} = -\frac{(-14.1\,\mathrm{mV}) - 14.1\,\mathrm{mV}}{1\,\mathrm{mV} - (-1\,\mathrm{mV})} = -14.1, \tag{2.103}$$

identical to the theoretical value, computed using Eq. (2.99). For illustrating the influence of the biasing current I_O on the value of the differential-mode voltage gain A_{DD}, a *DC sweep analysis* is performed in LTspice, for a variation range of the voltage v_{ID} between -1 mV and 1 mV, and also a parametrical simulation for $x = I_O$, between 10 µA and 40 µA, with a step of 10 µA. The circuit used for this simulation is presented in Fig. 2.50, and the simulation results are shown in Fig. 2.51. In order to directly view the results of the simulation, the directive ".save V(n010)" is introduced in the file, with V(n010) representing the differential-mode output voltage of the circuit v_{OD}.

For illustrating the influence of the load resistances R_1 and R_2 on the value of the differential-mode voltage gain A_{DD}, a *DC sweep analysis* is performed in LTspice, for a variation range of the voltage v_{ID} between -1 mV and 1 mV, and also a parametrical simulation for $x = R_1 = R_2$, between 60 kΩ and 120 kΩ, with a step of 20 kΩ. The circuit used for this simulation is presented in Fig. 2.52, and the simulation results are shown in Fig. 2.53. In order to directly view the result of the simulation, the directive ".save V(n010)" is introduced in the file, with V(n010) representing the differential-mode output voltage of the circuit v_{OD}.

2.3.5.3 Simulations for evaluating the maximal range of the common-mode input voltage, $\left[V_{IC}^{\mathrm{MIN}}, V_{IC}^{\mathrm{MAX}}\right]$

In order to determine the simulated value of the minimal common-mode input voltage V_{IC}^{MIN}, a *DC sweep analysis* is performed in LTspice, for a variation of the

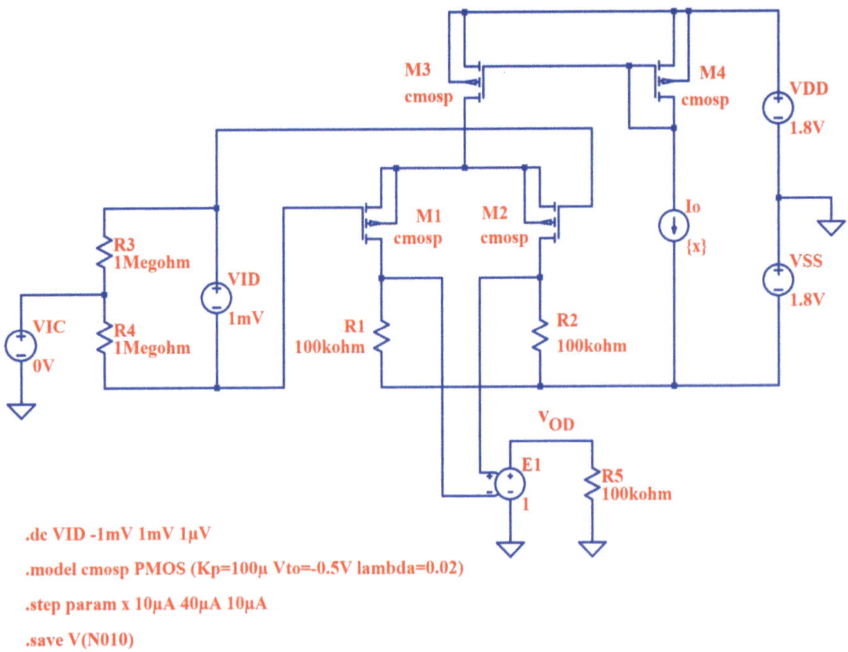

Fig. 2.50 The circuit used for simulating the small signal characteristics $v_{OD}(v_{ID})$, considering as parameter the biasing current I_O

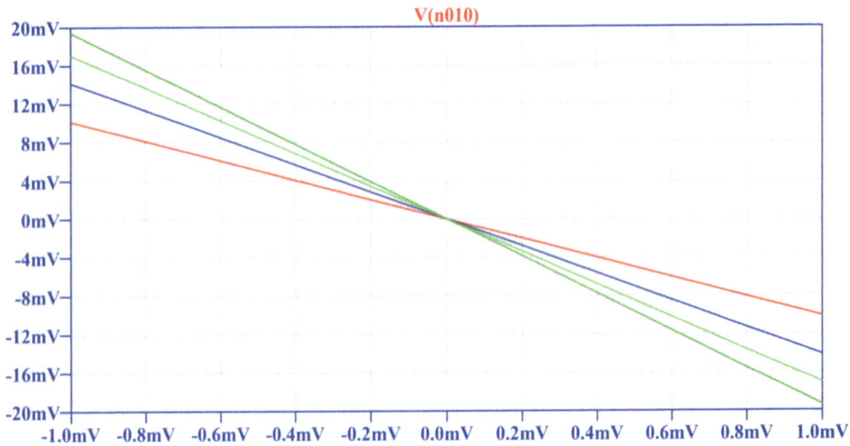

Fig. 2.51 Simulated small signal characteristics $v_{OD}(v_{ID})$, considering as parameter the biasing current I_O

voltage v_{ID} between $-1\,\text{mV}$ and $1\,\text{mV}$, and also a parametrical simulation for $x = V_{IC}$, with the parameter x having three values in the neighbourhood of the theoretical value $V_{IC}^{\text{MIN}} = -1.3\,\text{V}$ (Eq. (2.102)): $-1.26\,\text{V}$, $-1.28\,\text{V}$ and $-1.30\,\text{V}$.

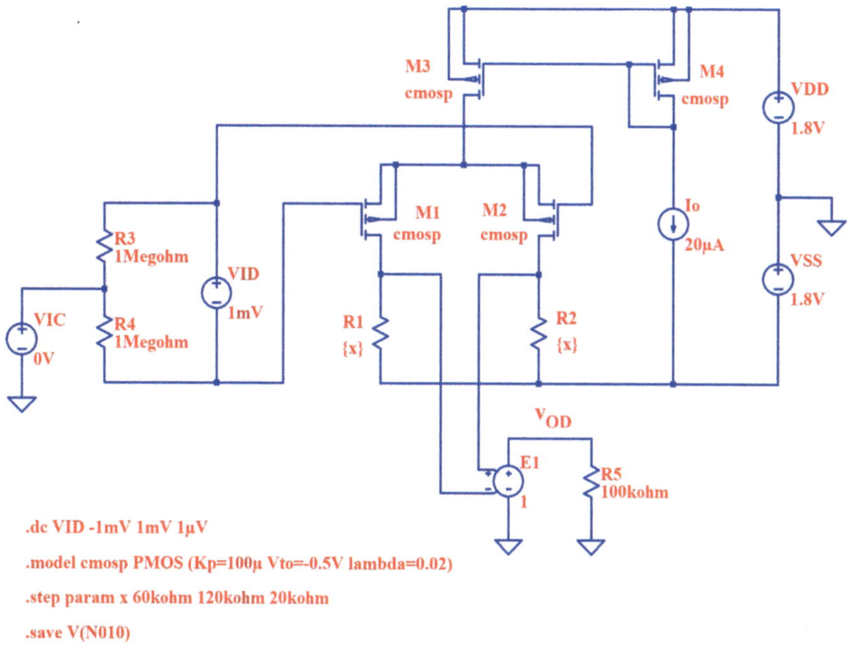

Fig. 2.52 The circuit used for simulating the small signal characteristics $v_{OD}(v_{ID})$, considering as parameter the load resistances $R_1 = R_2$

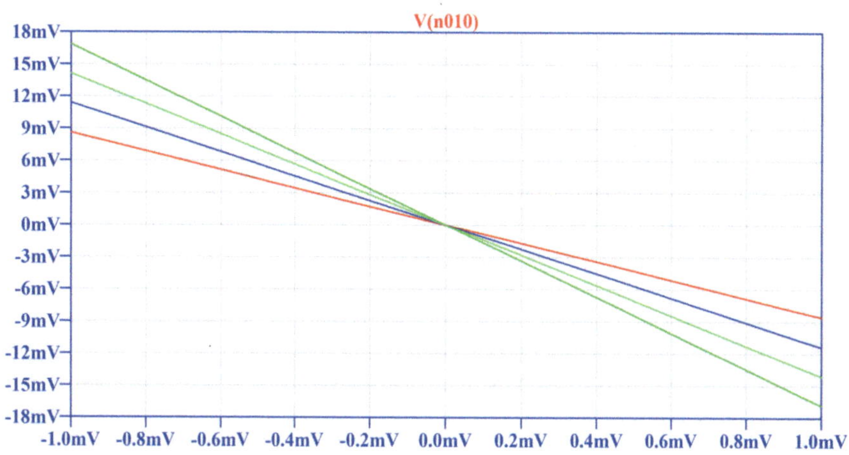

Fig. 2.53 Simulated small signal characteristics $v_{OD}(v_{ID})$, considering as parameter the load resistances $R_1 = R_2$

The circuit used for this simulation is presented in Fig. 2.54, and the simulation results are shown in Fig. 2.55. In order to directly view the results of the simulation, the directive ".save V(n010)" is introduced in the file, with V(n010) representing the differential-mode output voltage of the circuit v_{OD}.

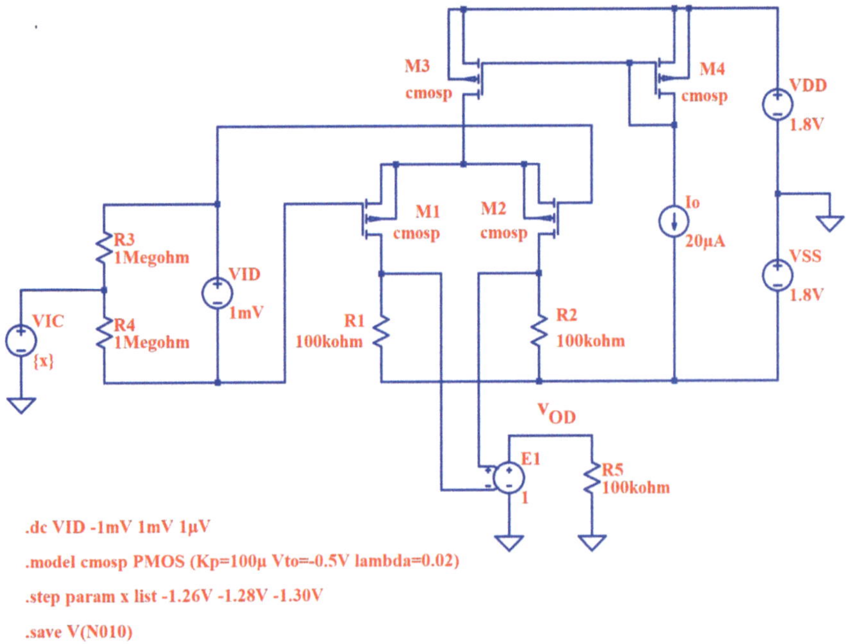

Fig. 2.54 Circuit for determining the simulated value of the minimal common-mode input voltage V_{IC}^{MIN}—parametrical simulation $v_{OD}(v_{ID})$

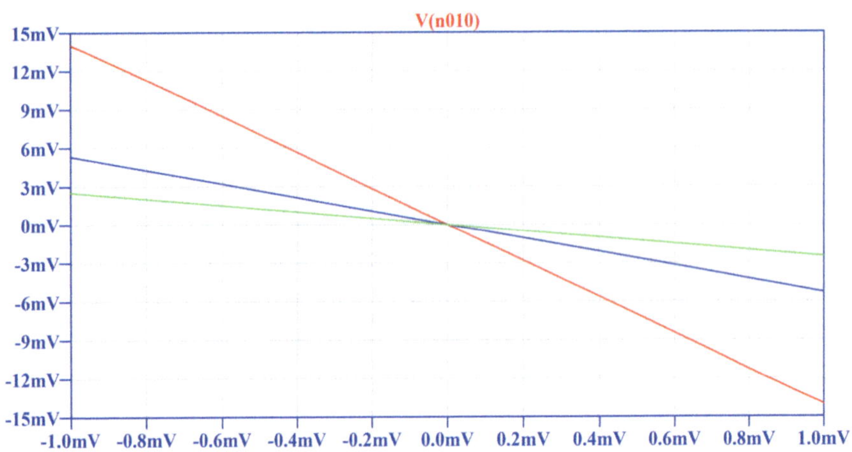

Fig. 2.55 Simulation results for determining the simulated value of the minimal common-mode input voltage V_{IC}^{MIN}—parametrical simulation $v_{OD}(v_{ID})$

The simulated value of V_{IC}^{MIN} is approximately $-1.26\,\text{V}$, very close to the theoretical value expressed by Eq. (2.102).

For applications that require an operation of the PMOS differential amplifier with passive load at small values of the common-mode input voltage V_{IC}, close to its negative supply voltage $V_{SS} = -1.8\,\text{V}$, two possible methods for decreasing the value of the minimal common-mode input voltage V_{IC}^{MIN} can be used. Analysing the mathematical expression (2.95) of V_{IC}^{MIN}, these methods are as follows:

First Method: The Decreasing of V_{IC}^{MIN} By Reducing the Load Resistance R_1

For illustrating this method, the resistance R_1 is reduced from the value of 100 kΩ to the value 50 kΩ, maintaining the biasing current to the initial value of 20 μA. The new value of the common-mode input voltage V_{IC}^{MIN} can be determined using Eq. (2.95):

$$V_{IC}^{MIN} = -1.8\,\text{V} + \frac{20\,\mu\text{A}}{2} \times 50\,\text{k}\Omega - 0.5\,\text{V} = -1.8\,\text{V}. \tag{2.104}$$

The cost of decreasing the minimal common-mode input voltage V_{IC}^{MIN} from $-1.3\,\text{V}$ to $-1.8\,\text{V}$ using this design method is a reduction in the differential-mode voltage gain $A_{DD} = -g_{m1}(R_1//r_{ds1})$ (Eq. (2.43)), in the hypothesis of a differential-mode output voltage. Because the values of g_{m1} and r_{ds1} remain the same when changing the value of the load resistance R_1, the new value of the differential-mode voltage gain A_{DD} will be

$$A_{DD} = -141.4\,\mu\text{A/V} \times (50\,\text{k}\Omega//5\,\text{M}\Omega) \cong -7, \tag{2.105}$$

in comparison to the previous value $A_{DD} \cong -14.1$, determined using Eq. (2.99).

In order to determine the simulated value of the new minimal common-mode input voltage V_{IC}^{MIN}, a *DC sweep analysis* is performed in LTspice, for a variation of the voltage v_{ID} between -1 mV and 1 mV, and also a parametrical simulation for $x = V_{IC}$, with the parameter x having three values in the neighbourhood of the new theoretical value $V_{IC}^{MIN} = -1.8\,\text{V}$: $-1.76\,\text{V}$, $-1.78\,\text{V}$ and $-1.80\,\text{V}$. The circuit used for this simulation is presented in Fig. 2.56, and the simulation results are shown in Fig. 2.57. In order to directly view the results of the simulation, the directive ".save V(n010)" is introduced in the file, with V(n010) representing the differential-mode output voltage of the circuit v_{OD}.

The simulated value of the minimal common-mode input voltage V_{IC} is approximately $-1.76\,\text{V}$, very close to the new theoretical value, determined using Eq. (2.104).

The maximal absolute value of the slope for the characteristics $v_{OD}(v_{ID})$ from Fig. 2.57 corresponds to a common-mode input voltage still in the maximal range $\left[V_{IC}^{MIN}, V_{IC}^{MAX}\right]$. The value of 7 for this maximal absolute value of the slope is equal to

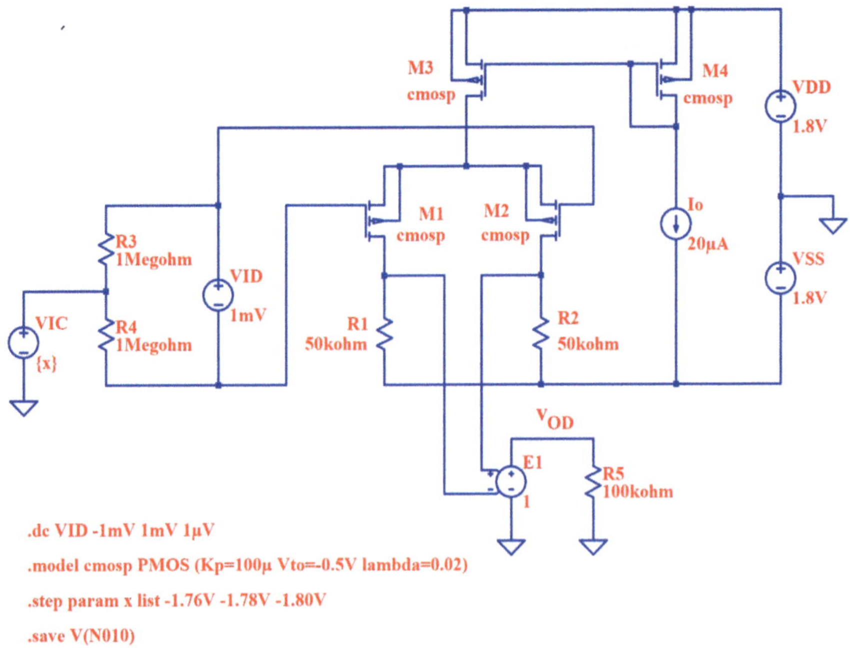

Fig. 2.56 Circuit for determining the simulated value of the decreased minimal common-mode input voltage V_{IC}^{MIN} using the first method—parametrical simulation $v_{OD}(v_{ID})$

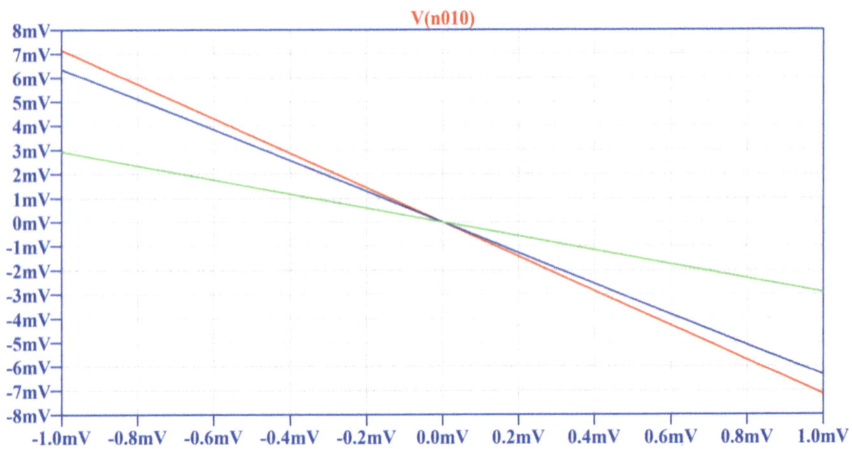

Fig. 2.57 Simulation results for determining the simulated value of the decreased minimal common-mode input voltage V_{IC}^{MIN} using the first method—parametrical simulation $v_{OD}(v_{ID})$

the new reduced absolute value of the differential-mode voltage gain A_{DD}, determined using Eq. (2.105). The other two characteristics from Fig. 2.57, having absolute values of their slopes inferior to the value of 7, correspond to a common-mode input voltage V_{IC} outside the interval $\left[V_{IC}^{\text{MIN}}, V_{IC}^{\text{MAX}}\right]$ ($V_{IC} = -1.78\,\text{V}$ and $V_{IC} = -1.80\,\text{V}$, respectively), equivalent to the transistors M1 and M2 no longer biased in saturation region, because $V_{IC} < V_{IC}^{\text{MIN}}$ for these values of V_{IC}.

Second Method: The Decreasing of V_{IC}^{MIN} By Reducing the Biasing Current I_O

For illustrating this method, the biasing current I_O is reduced from the value of 20 μA to the value 10 μA, maintaining the load resistance R_1 at the initial value of 100 kΩ. The new value of the common-mode input voltage V_{IC}^{MIN} can be determined using Eq. (2.95):

$$V_{IC}^{\text{MIN}} = -1.8\text{V} - \frac{10\,\mu\text{A}}{2} \times 100\,\text{k}\Omega + 0.5\,\text{V} = -1.8\,\text{V}. \tag{2.106}$$

The cost of decreasing the minimal common-mode input voltage V_{IC}^{MIN} from $-1.3\,\text{V}$ to $-1.8\,\text{V}$ is a reduction in the differential-mode voltage gain $A_{DD} = -g_{m1}(R_1//r_{ds1})$ (Eq. (2.43)), in the hypothesis of a differential-mode output voltage.

In order to determine the new theoretical value of the differential-mode voltage gain expressed using Eq. (2.43), it is necessary to compute the new values of transconductance g_{m1} and of the internal resistance r_{ds1} of transistor M1:

$$g_{m1} = \sqrt{2 \times 100\,\mu\text{A/V}^2 \times 10 \times 5\,\mu\text{A}} = 100\,\mu\text{A/V}, \tag{2.107}$$

$$r_{ds1} = \frac{1}{0.02\,\text{V}^{-1} \times 5\,\mu\text{A}} = 10\,\text{M}\Omega. \tag{2.108}$$

The new theoretical value of A_{DD} will be

$$A_{DD} = -100\,\mu\text{A/V} \times (100\,\text{k}\Omega//10\,\text{M}\Omega) \cong -10, \tag{2.109}$$

in comparison to with the previous value $A_{DD} \cong -14.1$, determined using Eq. (2.99).

In order to determine the simulated value of the new minimal common-mode input voltage V_{IC}^{MIN}, a *DC sweep analysis* is performed in LTspice, for a variation of the voltage v_{ID} between $-1\,\text{mV}$ and $1\,\text{mV}$, and also a parametrical simulation for $x = V_{IC}$, with the parameter x having three values in the neighbourhood of the new theoretical value $V_{IC}^{\text{MIN}} = -1.8\,\text{V}$: $-1.76\,\text{V}$, $-1.78\,\text{V}$ and $-1.80\,\text{V}$. The circuit used for this simulation is presented in Fig. 2.58, and the simulation results are shown in Fig. 2.59. In order to directly view the results of the simulation, the directive ".save

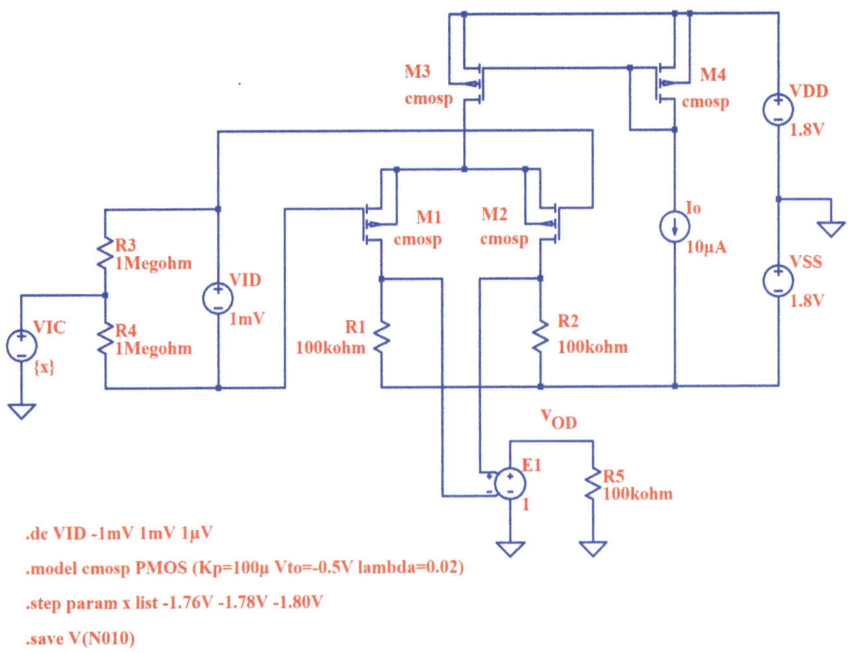

Fig. 2.58 Circuit for determining the simulated value of the decreased minimal common-mode input voltage V_{IC}^{MIN} using the second method—parametrical simulation $v_{OD}(v_{ID})$

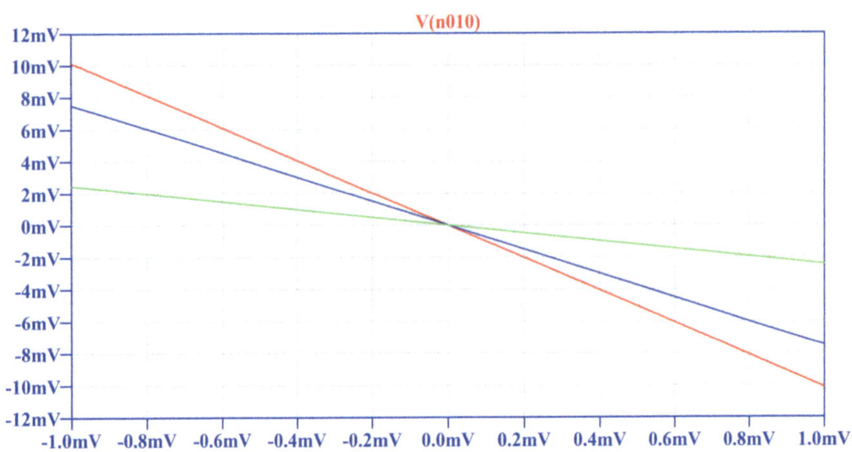

Fig. 2.59 Simulation results for determining the simulated value of the increased maximal common-mode input voltage V_{IC}^{MIN} using the second method—parametrical simulation $v_{OD}(v_{ID})$

V(n010)" is introduced in the file, with V(n010) representing the differential-mode output voltage of the circuit v_{OD}.

The simulated value of the minimal common-mode input voltage V_{IC} is approximately -1.76 V, very close to the theoretical value, determined using Eq. (2.104).

The maximal absolute value of the slope for the characteristics $v_{OD}(v_{ID})$ from Fig. 2.57 corresponds to a common-mode input voltage still in the maximal range $\left[V_{IC}^{\text{MIN}}, V_{IC}^{\text{MAX}}\right]$. The value of 10 for this maximal absolute value of the slope is equal to the new reduced absolute value of the differential-mode voltage gainA_{DD}, determined using Eq. (2.109). The other two characteristics from Fig. 2.59, having absolute values of their slopes inferior to the value of 10, correspond to a common-mode input voltage V_{IC} outside the interval $\left[V_{IC}^{\text{MIN}}, V_{IC}^{\text{MAX}}\right]$ ($V_{IC} = -1.78$ V and $V_{IC} = -1.80$ V, respectively), equivalent to the transistors M1 and M2 no longer biased in saturation region, because $V_{IC} < V_{IC}^{\text{MIN}}$ for these values of V_{IC}.

In conclusion, in order to decrease the minimal value of the common-mode input voltage V_{IC}^{MIN}, the second method is better than the first method. Its cost related to the reduction in the differential-mode voltage gain A_{DD} is smaller (from -14.1 to -10), compared with the reduction from -14.1 to -7 for the first method, both methods having the same benefit of decreasing V_{IC}^{MIN} from -1.3 V to -1.80 V.

In order to determine the simulated value of the maximal common-mode input voltage V_{IC}^{MAX}, a *DC sweep analysis* is performed in LTspice, for a variation of the voltage v_{ID} between -1 mV and 1 mV, and also a parametrical simulation for $x = V_{IC}$, with the parameter x having three values in the neighbourhood of the theoretical value $V_{IC}^{\text{MAX}} = 0.96$ V (Eq. (2.101)): 1.01 V, 1.03 V and 1.05 V. The circuit used for this simulation is presented in Fig. 2.60, and the simulation results are shown in Fig. 2.61. In order to directly view the results of the simulation, the directive ".save V(n010)" is introduced in the file, with V(n010) representing the differential-mode output voltage of the circuit v_{OD}.

The simulated value of V_{IC}^{MAX} is approximately 1.01 V, close to the theoretical value expressed by Eq. (2.101).

2.3.5.4 Simulations for Evaluating the Common-Mode Voltage Gain A_{CC}

In order to determine the common-mode voltage gain A_{CC} of the basic PMOS passive load differential amplifier from Fig. 2.39, a *DC sweep analysis* is performed in LTspice, for a variation of the common-mode input voltage V_{IC} between -0.5 V and 0.5 V. This interval $[-0.5\text{ V}, 0.5\text{ V}]$ is set for a proper operation of the differential amplifier, so it must be included in the maximal interval defined by $\left[V_{IC}^{\text{MIN}}, V_{IC}^{\text{MAX}}\right]$. The circuit used for this simulation is presented in Fig. 2.62, and the simulation results are shown in Fig. 2.63. In order to directly view the result of the simulation, the directive ".save V(n005)" is introduced in the file, with V(n005) representing the common-mode output voltage of the circuit v_{OC}.

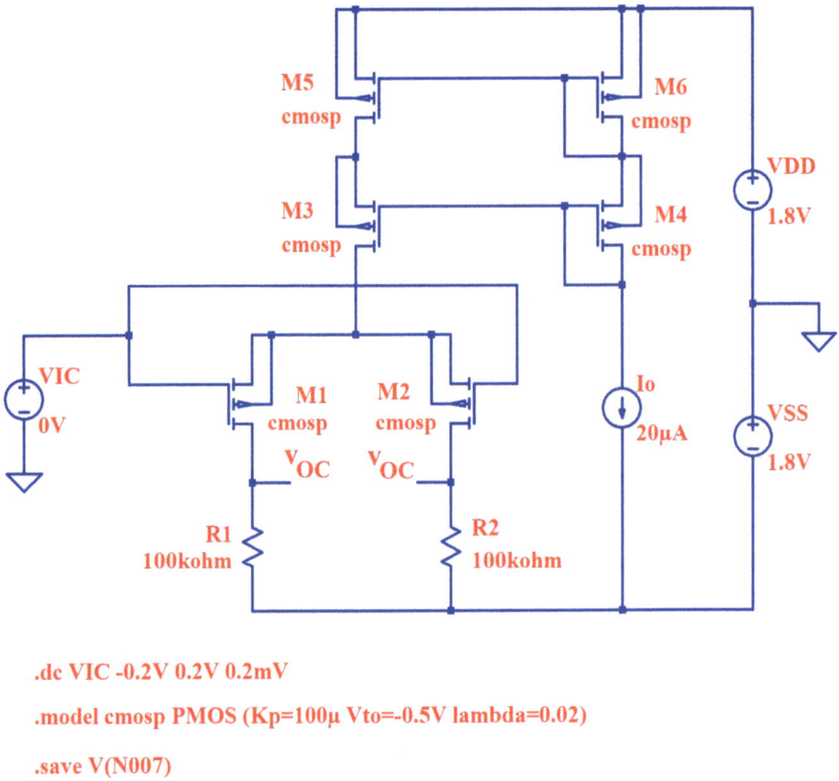

Fig. 2.66 The circuit used for simulating the dependence $v_{OC}(v_{IC})$, when biasing the PMOS differential amplifier using a cascode current mirror

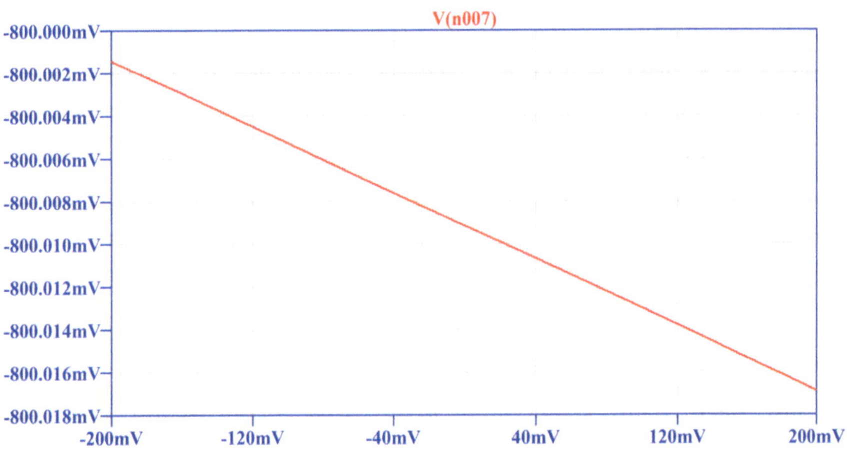

Fig. 2.67 Simulated characteristic $v_{OC}(v_{IC})$, when biasing the PMOS differential amplifier using a cascode current mirror

the expression of the common-mode voltage gain for the differential amplifier biased at a cascode current mirror from Fig. 2.66 can be expressed as follows:

$$A_{CC} \cong -\frac{R_1}{2g_{m3}r_{ds3}^2}. \tag{2.112}$$

The value of transconductance g_{m3} is:

$$g_{m3} = \sqrt{2 \times 100\,\mu\text{A/V}^2 \times 10 \times 20\,\mu\text{A}} = 200\,\mu\text{A/V} \tag{2.113}$$

and the value of r_{ds3} is:

$$r_{ds3} = \frac{1}{0.02\,\text{V}^{-1} \times 20\,\mu\text{A}} = 2.5\,\text{M}\Omega, \tag{2.114}$$

resulting in the following theoretical value of A_{CC}:

$$A_{CC} = -\frac{100\,\text{k}\Omega}{2 \times 200\,\mu\text{A/V} \times 6.25(\text{M}\Omega)^2} = -4 \times 10^{-5}, \tag{2.115}$$

close to the simulated value of -3.9×10^{-5}, obtained using Eq. (2.111).

In order to evaluate the dependence of the common-mode voltage gain for the differential amplifier biased at a cascode current mirror on the technological parameter λ_n, which models the channel-length modulation effect, an additional parametrical simulation for $x = \lambda_n$, between $0.02\,\text{V}^{-1}$ and $0.08\,\text{V}^{-1}$, with a step of $0.02\,\text{V}^{-1}$ is added to the previous *DC sweep analysis*, for a variation of the common-mode input voltage V_{IC} between $-0.2\,\text{V}$ and $0.2\,\text{V}$. The circuit used for this simulation is presented in Fig. 2.68, and the simulation results are shown in Fig. 2.69. In order to directly view the result of the simulation, the directive ".save V(n007)" is introduced in the file, with V(n007) representing the common-mode output voltage of the circuit v_{OC}.

The absolute value of the common-mode voltage gain A_{CC} (which is equal to the absolute value of the slope of the characteristics shown in Fig. 2.69) increases when the channel-length modulation parameter increases, from $0.01\,\text{V}^{-1}$ to $0.05\,\text{V}^{-1}$.

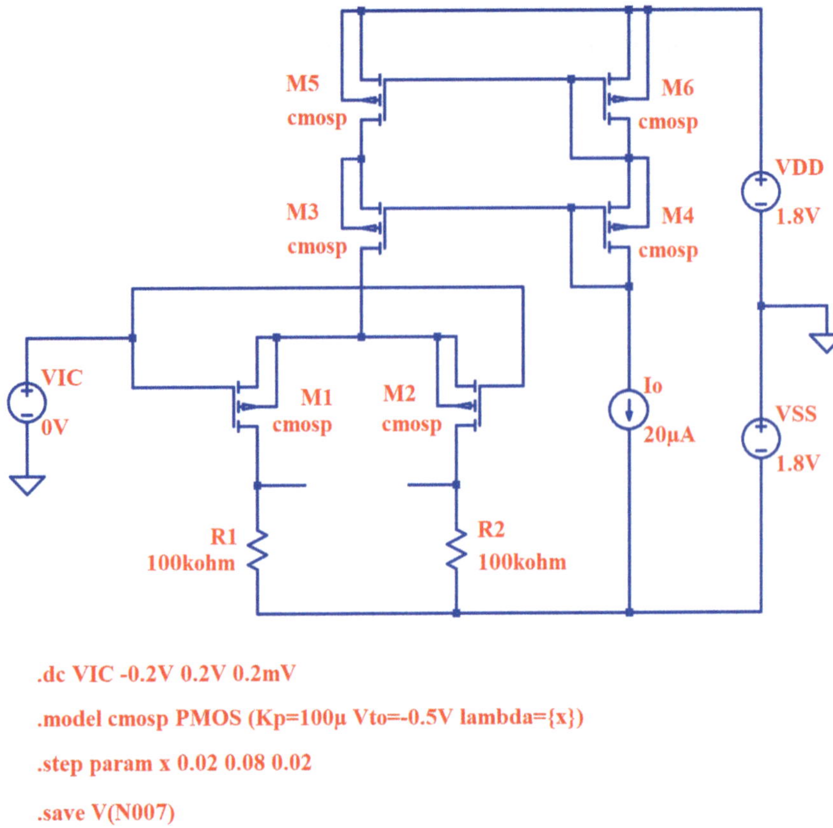

Fig. 2.68 Circuit used for simulating the dependence $v_{OC}(v_{IC})$, with an additional parametrical simulation for $x = \lambda_n$ (for the PMOS differential amplifier biased at a cascode current mirror)

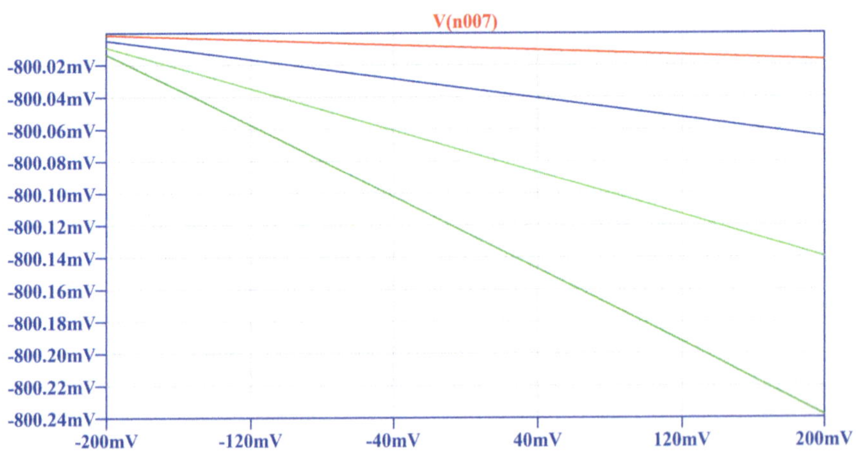

Fig. 2.69 Simulated characteristic $v_{OC}(v_{IC})$, with an additional parametrical simulation for $x = \lambda_n$ (for the PMOS differential amplifier biased at a cascode current mirror)

2.4 Comparison Between the Basic NMOS Differential Amplifier with Passive Load and the Basic PMOS Differential Amplifier with Passive Load

A comparison between the previous circuits (basic NMOS differential amplifier with passive load and basic PMOS differential amplifier with passive load) is presented in Table 2.1, from the point of view of their most important common-mode parameters:

- Minimal common-mode input voltage V_{IC}^{MIN} for the biasing of both differential amplifiers using a basic current mirror
- Maximal common-mode input voltage V_{IC}^{MAX} for the biasing of both differential amplifiers using a basic current mirror
- Common-mode voltage gain for a biasing of both differential amplifiers using a basic current mirror $A_{CC-BASIC}$
- Common-mode voltage gain for a biasing of both differential amplifiers using a cascode current mirror $A_{CC-CASCODE}$
- Differential-mode voltage gain A_{DD}

The results are obtained for a symmetrical supply voltage $V_{DD} = 1.8\,\text{V}$ and $V_{SS} = -1.8\,\text{V}$.

The most important conclusions of this comparison are the following:

- The basic NMOS differential amplifier with passive load can properly operate at large values of the common-mode input voltage V_{IC}, even greater than the positive supply voltage V_{DD}, and it does not work at extremely small values of the common-mode input voltage V_{IC}, close to V_{SS}, because the output voltage of the current mirror M3–M4 from Fig. 2.6 becomes smaller than its minimal output voltage $V_{O\min}$.
- The limitation of the minimal value of the common-mode input voltage V_{IC}^{MIN} becomes even more restrictive for a biasing of the basic NMOS differential amplifier with passive load using a cascode current mirror, because the cascode current mirror has a much greater minimal output voltage $V_{O\min}$ in comparison to the basic current mirror. This behavior is not illustrated in Table 2.1, but it can be verified by theoretical analysis and by LTspice simulations.
- For the basic NMOS differential amplifier with passive load, the benefit of biasing using a cascode current mirror is represented by an extremely important

Table 2.1 Comparison between basic NMOS differential amplifier with passive load and basic PMOS differential amplifier with passive load

Differential amplifier/parameter	Basic NMOS differential amplifier with passive load	Basic PMOS differential amplifier with passive load
Fig. no.	Fig. 2.6	Fig. 2.39
V_{IC}^{MIN}	−1.059 V	−1.8 V or less
V_{IC}^{MAX}	1.8 V or more	0.96 V
$A_{CC-BASIC}$	−0.01	−0.02
$A_{CC-CASCODE}$	-7.07×10^{-6}	-4×10^{-5}
A_{DD}	−20	−14.1

decreasing of the absolute value of the common-mode voltage gain, $|A_{CC-CASCODE}| << |A_{CC-BASIC}|$. So, a trade-off must be made between the optimization of the common-mode parameters: the value of the minimal common-mode input voltage V_{IC}^{MIN} or the common-mode voltage gain A_{CC}.

- The basic PMOS differential amplifier with passive load can properly operate at small values of the common-mode input voltage V_{IC}, even smaller than the negative supply voltage V_{SS}, and it does not work at extremely large values of the common-mode input voltage V_{IC}, close to V_{DD}, because the output voltage of the current mirror M3–M4 from Fig. 2.39 becomes smaller than its minimal output voltage $V_{O\,\mathrm{min}}$.
- The limitation of the maximal value of the common-mode input voltage V_{IC}^{MAX} becomes even more restrictive for a biasing of the basic NMOS differential amplifier with passive load using a cascode current mirror, because the cascode current mirror has a much greater minimal output voltage $V_{O\,\mathrm{min}}$ in comparison to the basic current mirror. This behavior is not illustrated in Table 2.1, but it can be verified by theoretical analysis and by LTspice simulations.
- For the basic PMOS differential amplifier with passive load, the benefit of biasing using a cascode current mirror is represented by an extremely important decreasing of the absolute value of the common-mode voltage gain, $|A_{CC-CASCODE}| << |A_{CC-BASIC}|$. So, a trade-off must be made between the optimization of the common-mode parameters: the value of the maximal common-mode input voltage V_{IC}^{MAX} or the common-mode voltage gain A_{CC}.
- The differences between the complementary NMOS and PMOS differential amplifiers with passive load related to their common-mode input voltage gains, $A_{CC-BASIC}$ and $A_{CC-CASCODE}$, are mainly given by the differences between the magnitudes of the channel-length modulation effects. This second-order effect is stronger for the PMOS transistors in comparison with the NMOS transistors ($\lambda_n = 0.01\,\mathrm{V}^{-1}$ and $\lambda_p = 0.02\,\mathrm{V}^{-1}$).
- The absolute value of the differential-mode voltage gain A_{DD} for the basic NMOS differential amplifier with passive load is superior to the absolute value of the same voltage gain of the basic PMOS differential amplifier with passive load, for the same biasing conditions, because the electron mobility is greater than hole mobility, which leads to a larger process transconductance parameter ($K'_n > K'_p$) and thus higher transconductance.

2.5 Basic NMOS Differential Amplifier with Active Load

2.5.1 *Small Signal Analysis*

The circuit of the basic NMOS differential amplifier with active load is presented in Fig. 2.70. The passive load from Fig. 2.6, represented by two identical physical resistors R_1 and R_2, is replaced in Fig. 2.70 with the current mirror M5–M6. The basic NMOS differential amplifier with active load has only a single output v_O.

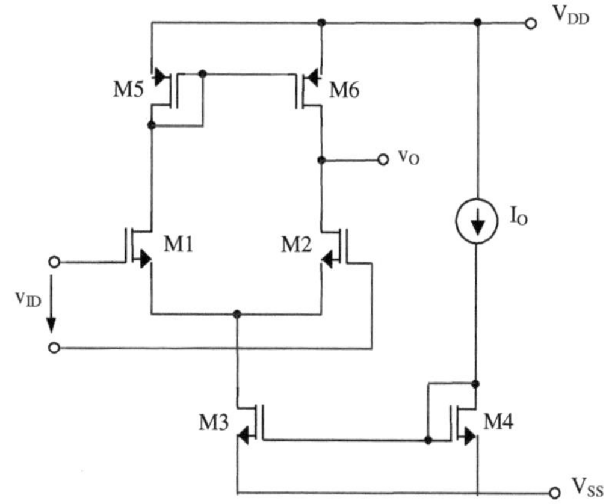

Fig. 2.70 Basic NMOS differential amplifier with active load

In order to determine the expression of the equivalent transconductance G_m of the basic MOS differential amplifier with active load, the simple small signal model of the MOS transistor must be considered (Fig. A.1 from Annex). Neglecting, in a first-order analysis, the drain-source resistances r_{ds} of MOS transistors, the small signal component of the drain current for transistor M1 is approximately $g_{m1}v_{GS1}$. Because, for a differential-mode analysis, the point of connecting the sources of transistors M1 and M2 has a zero potential (virtual ground point), the small signal component of the gate-source voltage of transistor M1 will be equal to $\frac{v_{ID}}{2}$. So, the small signal component of the drain current for transistor M1 becomes $g_{m1}\frac{v_{ID}}{2}$. Similarly, the small signal component of the drain current for transistor M2 will be $-g_{m2}\frac{v_{ID}}{2}$. Because of the current mirror M5–M6, the small signal component of the drain current for transistor M6 will be $g_{m1}\frac{v_{ID}}{2}$. The output current of the basic NMOS differential amplifier with active load will be the difference between small signal components of drain currents for transistors M6 and M2:

$$i_O = g_{m1}\frac{v_{ID}}{2} - \left(-g_{m2}\frac{v_{ID}}{2}\right). \tag{2.116}$$

Because transistors M1 and M2 are identical and they are biased in the DC operating point at the same drain current, their transconductances will be equal, $g_{m1} = g_{m2} \overset{not}{=} g_m$. So, Eq. (2.116) becomes

$$i_O = g_m v_{ID}. \tag{2.117}$$

In conclusion, the equivalent transconductance of the basic MOS differential amplifier with active load will be

$$G_m \overset{def}{=} \frac{i_O}{v_{ID}} = g_m = \sqrt{2K'_n\left(\frac{W}{L}\right)\frac{I_O}{2}} = \sqrt{K'_n I_O\left(\frac{W}{L}\right)}. \tag{2.118}$$

The output current i_O flows through the output resistance of the basic differential amplifier with active load, expressed by $r_{ds2}//r_{ds6}$, producing an output voltage equal to $i_O(r_{ds2}//r_{ds6})$. In consequence, the differential-mode voltage gain A_{DD} will have the following expression:

$$A_{DD} = \frac{v_O}{v_{ID}} = \frac{i_O(r_{ds2}//r_{ds6})}{v_{ID}} = \frac{g_m v_{ID}(r_{ds2}//r_{ds6})}{v_{ID}} = g_m(r_{ds2}//r_{ds6}). \tag{2.119}$$

The advantage of the basic NMOS differential amplifier with active load in comparison to the basic NMOS differential amplifier with passive load is related to the possibility of obtaining an integral differential-mode voltage gain A_{DD}, even for a single output (Eq. (2.119)). In comparison, the basic MOS differential amplifier with passive load has half the same value of A_{DD}, in the same conditions of a single output (Eq. (2.42)).

2.5.2 *Design Examples*

In order to confirm the previously presented theoretical estimated results, the basic MOS active load differential amplifier from Fig. 2.70 is simulated with LTspice using the following values: $K'_n = 200\,\mu\text{A/V}^2$, $K'_p = 100\,\mu\text{A/V}^2$, $\lambda_n = 0.01\,\text{V}^{-1}$, $\lambda_p = 0.02\,\text{V}^{-1}$, $V_{Tn} = 0.5\,\text{V}$, $V_{Tp} = -0.5\,\text{V}$, $V_{DD} = 1.8\,\text{V}$ and $V_{SS} = -1.8\,\text{V}$. MOS transistors have the aspect ratios $\left(\frac{W}{L}\right)_1 = \left(\frac{W}{L}\right)_2 = \left(\frac{W}{L}\right)_3 = \left(\frac{W}{L}\right)_4 = \frac{5\,\mu\text{m}}{0.5\,\mu\text{m}} = 10$. The value of the current source is $I_O = 20\,\mu\text{A}$, and the value of the output voltage source is $V_O = 1.159\,\text{V}$. The value of this voltage source is set for reducing the errors introduced by the channel-length modulation effect that affects the operation of the current mirror M5–M6. This particular value of the voltage source makes voltage $V_{SD6} = 1.8\,\text{V} - 1.159\,\text{V} = 0.641\,\text{V}$ equal to voltage V_{SD5}, expressed by

$$V_{SD5} = V_{SG5} = |V_{Tp}| + \sqrt{\frac{2(I_O/2)}{K'_p(W/L)}} = 0.5\,\text{V} + \sqrt{\frac{2 \times 10\,\mu\text{A}}{100\,\mu\text{A/V}^2 \times 10}}$$
$$= 0.641\,\text{V}. \tag{2.120}$$

The theoretical value of the equivalent transconductance G_m for the basic NMOS differential amplifier with active load is expressed by Eq. (2.118):

$$G_m = \sqrt{200\,\mu\text{A/V}^2 \times 20\,\mu\text{A} \times 10} \cong 200\,\mu\text{A/V}. \tag{2.121}$$

The maximal value of the common-mode input voltage is limited by the operation in saturation of both transistors M1 and M2:

$$V_{IC}^{MAX} = V_{DD} - V_{SG5} - V_{DS1sat} + V_{GS1}, \tag{2.122}$$

equivalent to

$$V_{IC}^{\mathrm{MAX}} = V_{DD} - (V_{SG5} - V_T). \tag{2.123}$$

So,

$$V_{IC}^{\mathrm{MAX}} = V_{DD} - \sqrt{\frac{I_O}{K'_p\left(\frac{W}{L}\right)}}, \tag{2.124}$$

resulting in

$$V_{IC}^{\mathrm{MAX}} = 1.8\,\mathrm{V} - \sqrt{\frac{20\,\mu\mathrm{A}}{100\,\mu\mathrm{A/V}^2 \times 10}} = 1.8\,\mathrm{V} - 0.141\,\mathrm{V} = 1.659\,\mathrm{V}. \tag{2.125}$$

The minimal value of the common-mode input voltage is limited by the biasing in saturation of the transistor M3:

$$V_{IC}^{\mathrm{MIN}} = -V_{SS} + V_{DS3sat} + V_{GS1}, \tag{2.126}$$

equivalent to

$$V_{IC}^{\mathrm{MIN}} = -V_{SS} + V_T + (V_{GS3} - V_T) + (V_{GS1} - V_T), \tag{2.127}$$

resulting in

$$V_{IC}^{\mathrm{MIN}} = -V_{SS} + V_T + \left(\sqrt{2} + 1\right)\sqrt{\frac{I_O}{K'_n\left(\frac{W}{L}\right)}}. \tag{2.128}$$

So,

$$V_{IC}^{\mathrm{MIN}} = -1.8\,\mathrm{V} + 0.5\,\mathrm{V} + 2.41 \times \sqrt{\frac{20\,\mu\mathrm{A}}{200\,\mu\mathrm{A/V}^2 \times 10}} = -1.059\,\mathrm{V}. \tag{2.129}$$

Simulations for evaluating the equivalent transconductance G_m

The circuit used for simulating the characteristic $i_O(v_{ID})$ for the basic NMOS active load differential amplifier is presented in Fig. 2.71, and the simulated characteristic $i_O(v_{ID})$ is shown in Fig. 2.72 (a *DC sweep analysis* is performed in LTspice, for a variation of the voltage v_{ID} between -1 mV and 1 mV, with a step of 1 μV). In order to directly view the result of the simulation, the directive ".save I(Vo)" is introduced in the file, with I(Vo) representing the output current of the circuit i_O.

From Fig. 2.72, the simulated value of the equivalent transconductance G_m for the basic MOS active load differential amplifier is

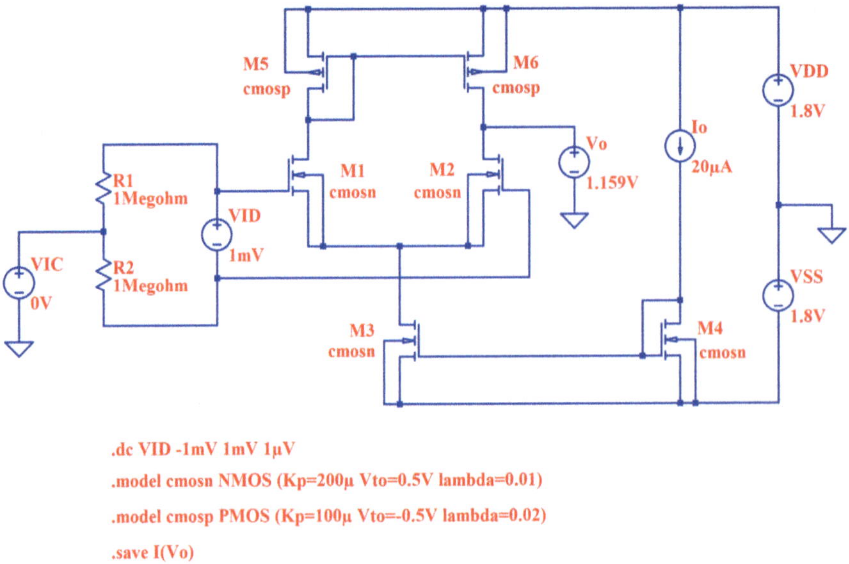

Fig. 2.71 Circuit used for simulating the characteristic $i_O(v_{ID})$ for the basic MOS active load differential amplifier

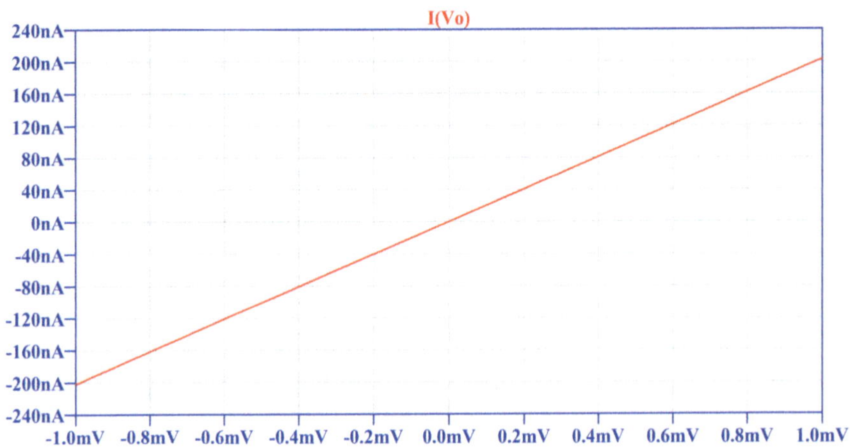

Fig. 2.72 Simulated characteristic $i_O(v_{ID})$ for the basic MOS active load differential amplifier

$$G_m = \frac{200\,\text{nA} - (-200\,\text{nA})}{1\,\text{mV} - (-1\,\text{mV})} = \frac{400\,\text{nA}}{2\,\text{mV}} = 200\,\mu\text{A/V}, \tag{2.130}$$

equal to the theoretical value, expressed by Eq. (2.121).

Simulations for evaluating the maximal range of the common-mode input voltage, $\left[V_{IC}^{MIN}, V_{IC}^{MAX}\right]$

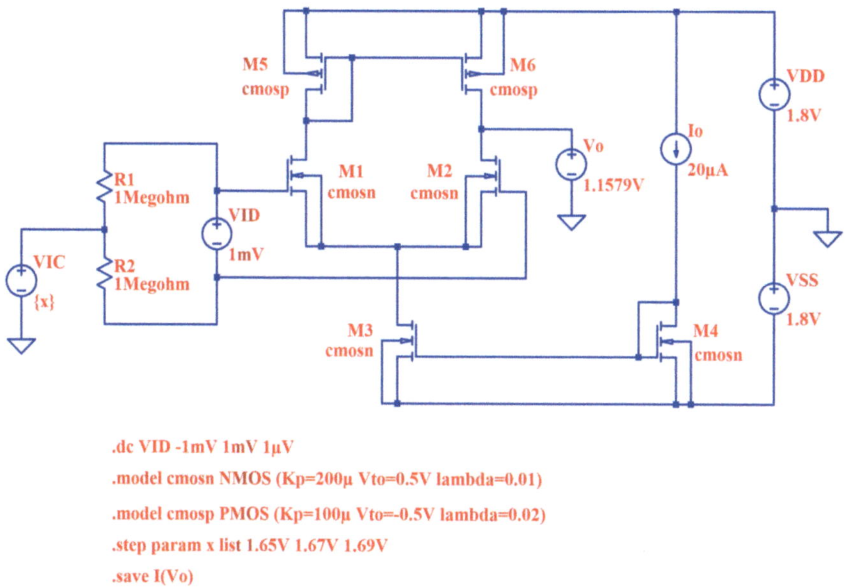

Fig. 2.73 Circuit for determining the simulated value of the maximal common-mode input voltage V_{IC}^{MAX}—parametrical simulation $i_O(v_{ID})$

In order to determine the simulated value of the maximal common-mode input voltage V_{IC}^{MAX}, a *DC sweep analysis* is performed in LTspice, for a variation of the voltage v_{ID} between -1 mV and 1 mV, and also a parametrical simulation for $x = V_{IC}$, with the parameter x having three values in the neighbourhood of the theoretical value $\mathrm{V_{IC}^{MAX}} = 1.659\,\mathrm{V}$ (Eq. (2.125)): 1.65 V, 1.67 V and 1.69 V. The circuit used for this simulation is presented in Fig. 2.73, and the simulation results are shown in Fig. 2.74. In order to directly view the results of the simulation, the directive ".save I(Vo)" is introduced in the file, with I(Vo) representing the output current of the circuit i_O.

The simulated value of V_{IC}^{MAX} is approximately 1.65 V, very close to the theoretical value expressed by Eq. (2.125).

The maximal slope of the characteristics $i_O(v_{ID})$ from Fig. 2.74 corresponds to a common-mode input voltage still in the maximal range $\left[V_{IC}^{MIN}, V_{IC}^{MAX}\right]$. The value of 200 µA/V of this maximal slope is equal to the theoretical value of the equivalent transconductance G_m, expressed by Eq. (2.121). The other two characteristics from Fig. 2.74, having the slopes inferior to the value of 200 µA/V, correspond to a common-mode input voltage V_{IC} outside the interval $\left[V_{IC}^{MIN}, V_{IC}^{MAX}\right]$ ($V_{IC} = 1.67$ V and $\mathrm{V_{IC}} = 1.69$ V, respectively), equivalent to the transistors M1 and M2 no longer biased in saturation region, because $V_{IC} > V_{IC}^{MAX}$ for these values of V_{IC}.

In order to determine the simulated value of the minimal common-mode input voltage V_{IC}^{MIN}, a *DC sweep analysis* is performed in LTspice, for a variation of the

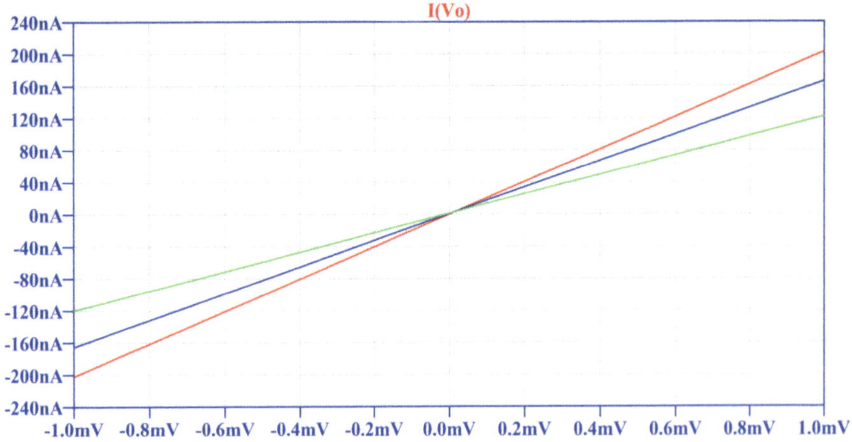

Fig. 2.74 Simulation results for determining the simulated value of the maximal common-mode input voltage V_{IC}^{MAX}—parametrical simulation $i_O(v_{ID})$

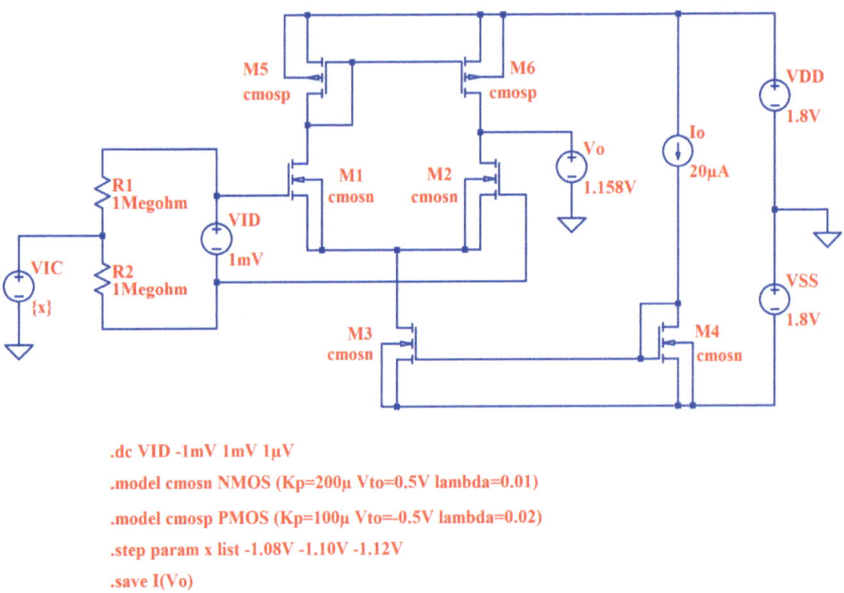

Fig. 2.75 Circuit for determining the simulated value of the minimal common-mode input voltage V_{IC}^{MIN}

voltage v_{ID} between $-1\,\text{mV}$ and $1\,\text{mV}$, and also a parametrical simulation for $x = V_{IC}$, with the parameter x having three values in the neighbourhood of the theoretical value $V_{IC}^{MIN} = -1.059\,\text{V}$ (Eq. (2.129)): $-1.08\,\text{V}$, $-1.10\,\text{V}$ and $-1.12\,\text{V}$. The circuit used for this simulation is presented in Fig. 2.75, and the simulation results are shown in Fig. 2.76. In order to directly view the results of the simulation,

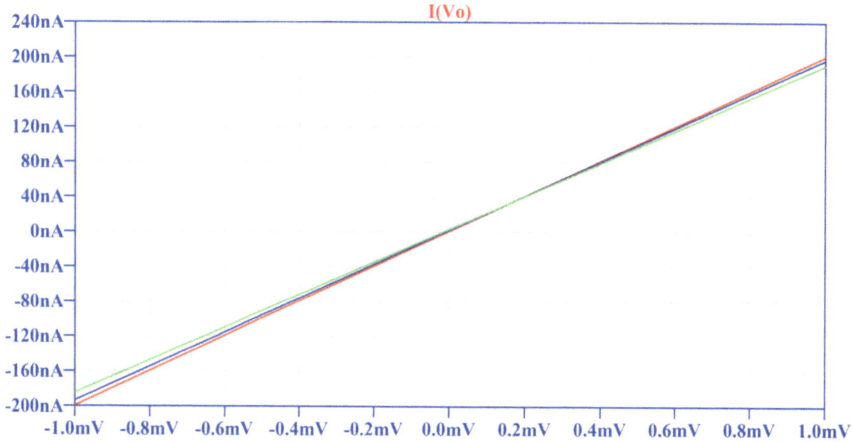

Fig. 2.76 Simulation results for determining the simulated value of the minimal common-mode input voltage V_{IC}^{MIN}

the directive ".save I(Vo)" is introduced in the file, with I(Vo) representing the output current of the circuit i_O.

The simulated value of V_{IC}^{MIN} is approximately $-1.08\,\mathrm{V}$, very close to the theoretical value expressed by Eq. (2.129).

The maximal slope of the characteristics $i_O(v_{ID})$ from Fig. 2.76 corresponds to a common-mode input voltage still in the maximal range $\left[V_{IC}^{\mathrm{MIN}}, V_{IC}^{\mathrm{MAX}}\right]$. The value of $200\,\mu\mathrm{A/V}$ of this maximal slope is equal to the theoretical value of the equivalent transconductance G_m, expressed by Eq. (2.121). The other two characteristics from Fig. 2.76, having the slopes inferior to the value of $200\,\mu\mathrm{A/V}$, correspond to a common-mode input voltage V_{IC} outside the interval $\left[V_{IC}^{\mathrm{MIN}}, V_{IC}^{\mathrm{MAX}}\right]$ ($V_{IC} = -1.10\,\mathrm{V}$ and $V_{IC} = -1.12\,\mathrm{V}$, respectively), equivalent to the transistor M3 no longer biased in saturation region, because $V_{IC} < V_{IC}^{\mathrm{MIN}}$ for these values of V_{IC}.

2.6 Differential Amplifier with Small Voltage Gain

2.6.1 Small Signal Analysis

In order to design a differential amplifier that can operate at high frequency, its differential-mode voltage gain A_{DD} must be small, having the order of magnitude of some unities. This voltage gain must be precisely controlled using a simple mathematical relation. The differential amplifier presented in Fig. 2.77 can be used for this purpose, with the decreasing of its differential-mode voltage gain being made by using a low-impedance equivalent load (transistors M5 and M6).

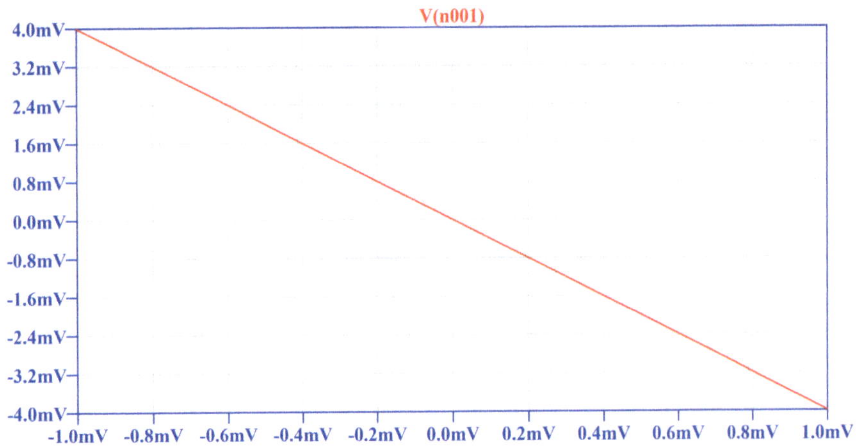

Fig. 2.80 Simulated characteristic $v_{OD}(v_{ID})$, for the differential amplifier with small voltage gain

From Fig. 2.80, the simulated value of the differential-mode voltage gain A_{DD} for the differential amplifier with small voltage gain is

$$A_{DD} = \frac{-4\,\text{mV} - 4\,\text{mV}}{1\,\text{mV} - (-1\,\text{mV})} = -\frac{8\,\text{mV}}{2\,\text{mV}} = -4, \tag{2.134}$$

identical to the theoretical value, expressed by Eq. (2.133).

2.7 Differential Amplifier with Small Controllable Voltage Gain

2.7.1 *Small Signal Analysis*

The differential amplifier presented in Fig. 2.81 can be used for a similar purpose with the circuit presented in Fig. 2.77, having the advantage of an additional possibility of controlling the differential-mode voltage gain A_{DD} by using two aspect ratios. Thus, the control of A_{DD} can be made by the aspect ratios $\frac{(W/L)_1}{(W/L)_5}$ and $\frac{(W/L)_7}{(W/L)_9}$.

Because of the circuit symmetry, the evaluation of A_{DD} can be made using the differential-mode half circuit, presented in Fig. 2.82.

The expression of the differential-mode voltage gain can be expressed as follows:

$$A_{DD} \cong -g_{m1}\left(\frac{1}{g_{m5}}\right) = -\sqrt{\frac{2K'_n\left(\frac{W}{L}\right)_1 I_{D1}}{2K'_p\left(\frac{W}{L}\right)_5 I_{D5}}}. \tag{2.135}$$

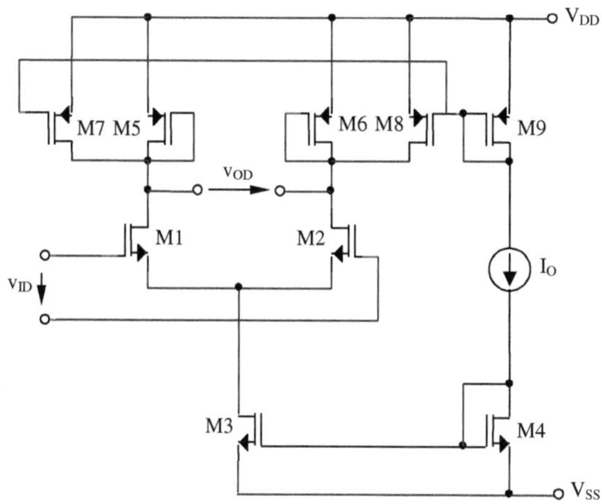

Fig. 2.81 Differential amplifier with small controllable voltage gain

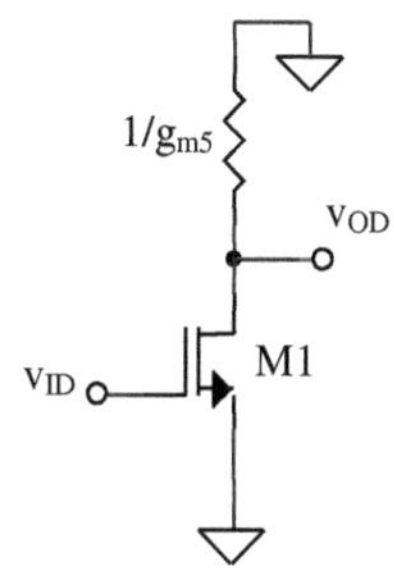

Fig. 2.82 Differential-mode half circuit for the differential amplifier with small controllable voltage gain

Because $I_{D1} = \frac{I_O}{2}$ and $I_{D5} = I_{D1} - I_{D7} = \frac{I_O}{2} - \frac{\left(\frac{W}{L}\right)_7}{\left(\frac{W}{L}\right)_9} I_O$, the previous relation becomes

$$A_{DD} \cong -g_{m1}\left(\frac{1}{g_{m5}}\right) = -\sqrt{\frac{K'_n\left(\frac{W}{L}\right)_1}{K'_p\left(\frac{W}{L}\right)_5}\frac{\left(\frac{W}{L}\right)_9}{\left(\frac{W}{L}\right)_9 - 2\left(\frac{W}{L}\right)_7}}. \tag{2.136}$$

2.7.2 Design Example

In order to confirm the previously presented theoretical estimated results, the differential amplifier with small controllable voltage gain from Fig. 2.81 is simulated with LTspice using the following values: $K'_n = 200\,\mu\text{A/V}^2$, $K'_p = 100\,\mu\text{A/V}^2$, $\lambda_n = 0.01\,\text{V}^{-1}$, $\lambda_p = 0.02\,\text{V}^{-1}$, $V_{Tn} = 0.5\,\text{V}$, $V_{Tp} = -0.5\,\text{V}$, $V_{DD} = 1.8\,\text{V}$ and $V_{SS} = -1.8\,\text{V}$. MOS transistors have the aspect ratios $\left(\frac{W}{L}\right)_1 = \left(\frac{W}{L}\right)_2 = \frac{20\,\mu\text{m}}{0.5\,\mu\text{m}} = 40$, $\left(\frac{W}{L}\right)_3 = \left(\frac{W}{L}\right)_4 = \frac{5\,\mu\text{m}}{0.5\,\mu\text{m}} = 10$, $\left(\frac{W}{L}\right)_5 = \left(\frac{W}{L}\right)_6 = \frac{5\,\mu\text{m}}{1\,\mu\text{m}} = 5$, $\left(\frac{W}{L}\right)_7 = \left(\frac{W}{L}\right)_8 = \frac{1\,\mu\text{m}}{0.5\,\mu\text{m}} = 2$ and $\left(\frac{W}{L}\right)_9 = \frac{5\,\mu\text{m}}{0.5\mu\text{m}} = 10$. The value of the current source is $I_O = 20\,\mu\text{A}$.

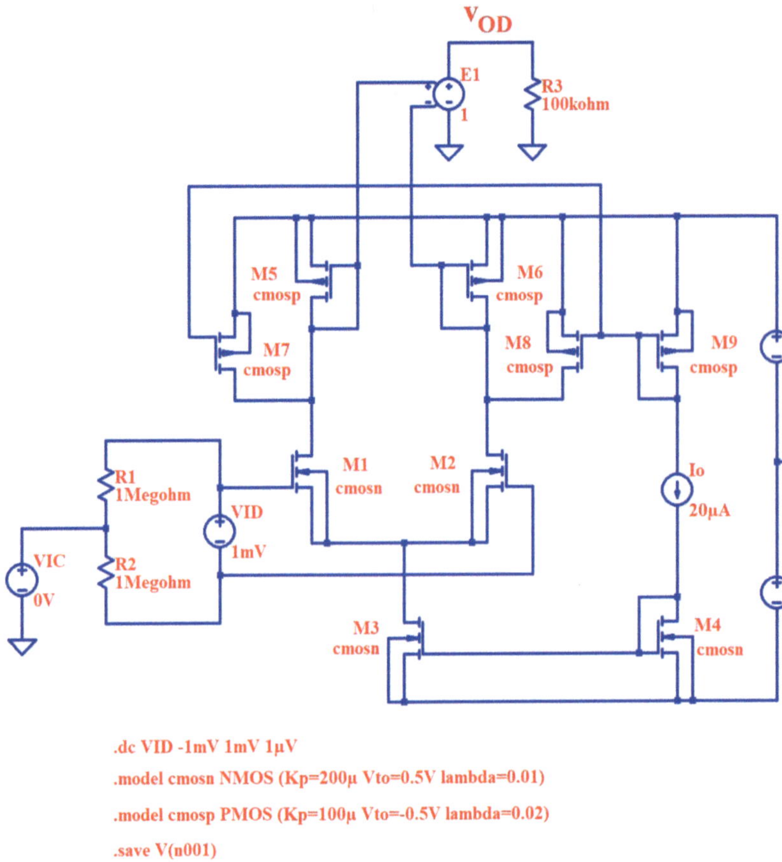

Fig. 2.83 Circuit used for simulating the characteristic $v_{OD}(v_{ID})$, for the differential amplifier with small controllable voltage gain

The theoretical value of A_{DD} will be

$$A_{DD} \cong -\sqrt{\frac{200\,\mu A/V^2 \times 40}{100\,\mu A/V^2 \times 5}\,\frac{10}{10-2\times 2}} = -5.16. \tag{2.137}$$

The circuit used for simulating the characteristic $v_{OD}(v_{ID})$ for the differential amplifier with small controllable voltage gain is presented in Fig. 2.83. The simulated characteristic $v_{OD}(v_{ID})$ for the differential amplifier with small controllable voltage gain is shown in Fig. 2.84 (a *DC sweep analysis* is performed in LTspice, for a variation of the voltage v_{ID} between -1 mV and 1 mV, with a step of 1 μV). In order to directly view the result of the simulation, the directive ".save V(n001)" is introduced in the file, with V(n001) representing the differential-mode output voltage of the circuit v_{OD}.

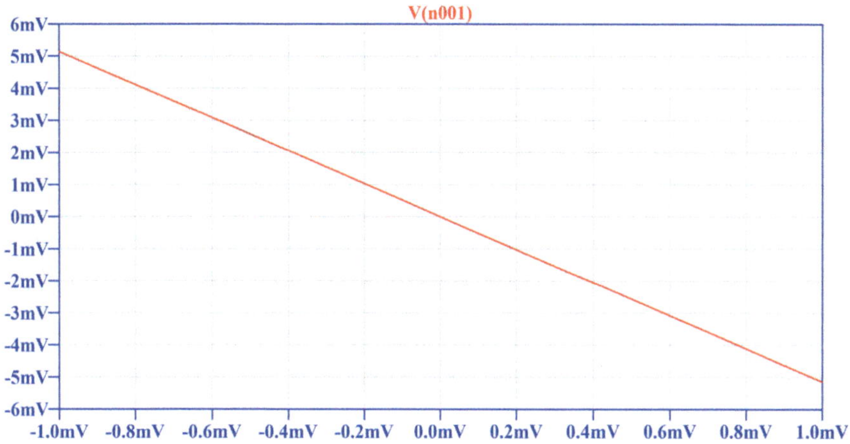

Fig. 2.84 Simulated characteristic $v_{OD}(v_{ID})$, for the differential amplifier with small controllable voltage gain

From Fig. 2.84, the simulated value of the differential-mode voltage gain A_{DD} for the differential amplifier with small controllable voltage gain is

$$A_{DD} = \frac{-5.14\,\text{mV} - 5.14\,\text{mV}}{1\,\text{mV} - (-1\,\text{mV})} = -\frac{10.28\,\text{mV}}{2\,\text{mV}} = -5.14, \tag{2.138}$$

very close to the theoretical value, expressed by Eq. (2.137).

References

1. Razavi B (2001) Design of Analog CMOS Integrated Circuits. New York: McGraw-Hill Education
2. Gray PR, Hurst PJ, Lewis SH and Meyer RG (2009) Analysis and Design of Analog Integrated Circuits, 5th ed. New York: Wiley
3. Allen PE and Holberg DR (2011) CMOS Analog Circuit Design, 3rd ed. New York: Oxford University Press
4. Sedra AS and Smith KC (2015) Microelectronic Circuits, 7th ed. New York: Oxford University Press

Chapter 3
CMOS Operational Amplifier Architectures: From Basic to Cascode and Folded Cascode Structures

Abstract Operational amplifiers are among the most versatile and widely used building blocks in analog integrated circuits. Their internal structure is, however, composed of relatively simple stages such as differential pairs, current mirrors and gain stages. A clear understanding of these internal architectures is essential for interpreting their performance, limitations and proper application in practical analog design. This chapter provides a systematic analysis of several fundamental CMOS operational amplifier architectures, including basic two-stage structures, cascode configurations and folded cascode topologies. The study is conducted using a simplified MOS transistor model based on the quadratic current-voltage relationship in saturation, with channel-length modulation included as the only second-order effect. For each architecture, the operating principle is first described in detail, followed by the determination of the DC operating point currents and the derivation of key small-signal performance parameters. These include voltage gain, input and output resistance and allowable common-mode input voltage range. The analysis emphasizes compact analytical expressions that offer direct physical insight into circuit operation and trade-offs. Unlike the preceding chapters, this chapter is primarily theoretical in nature, with the goal of building a solid conceptual foundation for understanding operational amplifier behavior. This foundation supports the more application-oriented analyses presented in the following chapter.

Keywords Operational amplifier architecture · Voltage gain · Input and output resistance · Maximal range of the common-mode input voltage · Cascode architecture · Folded cascode architecture

3.1 Introduction

Operational amplifiers are among the most versatile and widely used components in analog integrated circuits. Their internal structure (extensively described in the classical analog integrated circuit design literature [1–4]), however, is based on a combination of relatively simple building blocks, such as differential amplifiers, current mirrors and gain stages. Understanding these internal architectures is

C. R. Popa, *Fundamentals of CMOS Analog Integrated Circuits*,
https://doi.org/10.1007/978-3-032-30677-7_3

essential for gaining insight into their performance, limitations and appropriate use in practical applications.

This chapter presents a systematic study of several fundamental CMOS operational amplifier architectures, including basic structures, cascode configurations and folded cascode topologies. The analysis is carried out using a simplified MOS transistor model based on the quadratic current-voltage characteristic in saturation, including channel-length modulation as the only second-order effect.

For each architecture, the operating principle is explained in detail, followed by the calculation of the drain currents in the DC operating point and the derivation of key performance parameters. These include voltage gain, input and output resistances and allowable common-mode input voltage range. The emphasis is placed on developing clear and tractable analytical expressions that provide direct insight into circuit behavior.

Unlike the other chapters, the focus here is primarily theoretical, aiming to establish a solid understanding of the internal functioning of the operational amplifiers. This knowledge forms the basis for the application-oriented analysis presented in the following chapter.

3.2 Two-Stage CMOS Operational Amplifier (Architecture 1 of a CMOS Operational Amplifier)

The two-stage operational amplifier topology employed in this paragraph follows standard CMOS design methodologies [1–4]. This circuit is presented in Fig. 3.1.

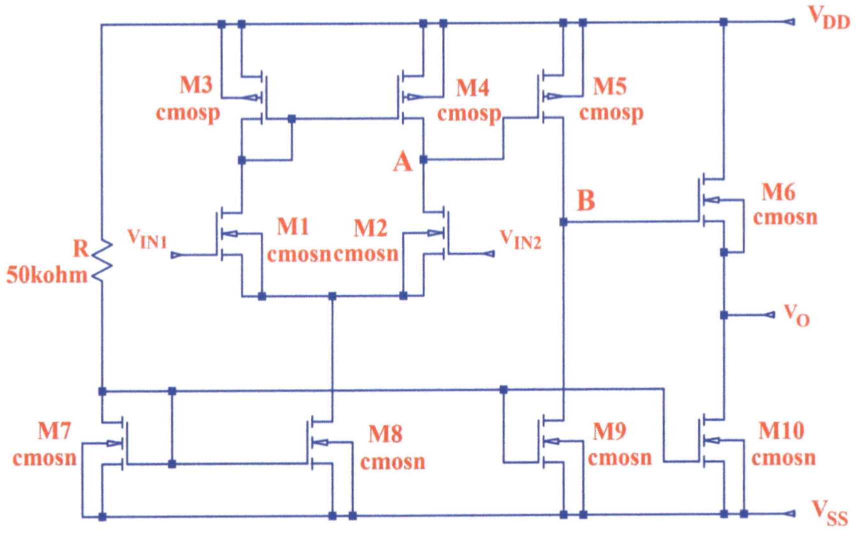

Fig. 3.1 Two-stage CMOS operational amplifier architecture

In order to evaluate the performance of the two-stage CMOS operational amplifier, the following parameters must be evaluated.

3.2.1 DC Operating Point Currents

Noting with I_{REF} the current passing through the resistor R, all DC operating point currents of the operational amplifier will linearly depend on the reference current I_{REF}. For simplicity, the following notation is used for the threshold voltages on NMOS and PMOS transistors: $V_T \overset{not}{=} V_{Tn} = -V_{Tp}$. In order to compute this current, the following equation can be written as

$$V_{DD} - V_{SS} = RI_{REF} + V_{GS7}, \tag{3.1}$$

with the reference current I_{REF} being expressed as a function of the gate-source voltage of transistor M7 using the squaring mathematical function, specific to a biasing in saturation region:

$$I_{REF} \cong \frac{K'_n}{2}\left(\frac{W}{L}\right)(V_{GS7} - V_T)^2. \tag{3.2}$$

The channel-length modulation is neglected in manual analysis. Replacing Eq. (3.2) in Eq. (3.1) and using symmetric supply voltages, $V_{SS} = -V_{DD}$, result in

$$2V_{DD} = R\frac{K'_n}{2}\left(\frac{W}{L}\right)(V_{GS7} - V_T)^2 + V_{GS7}, \tag{3.3}$$

which represents a second-order equation, having as unknown value the gate-source voltage of transistor M7, V_{GS7}. This equation has the solutions $(V_{GS7})_A$ and $(V_{GS7})_B$, with $(V_{GS7})_A > V_T$ and $(V_{GS7})_B < V_T$. It results that $(V_{GS7})_A$ represents the mathematical solution that fulfils the condition of biasing in saturation region of transistor M7. So, replacing $(V_{GS7})_A$ in Eq. (3.2), it is possible to obtain the expression of the reference current, I_{REF}.

The expressions of the DC operating point currents of all transistors from Fig. 3.1 as a function of the reference current I_{REF} are shown in Table 3.1. All transistors are set to be identical (having the same aspect ratio W/L), except transistor M9, which has an aspect ratio half of all the other aspect ratios. The reason is the necessity of reducing the errors introduced in the operation of the differential amplifier M1–M2 by the channel-length modulation effect. This can be achieved by ensuring an operation of transistors M1 and M2 at approximately the same drain-source voltages, which is equivalent to have approximately the same source-gate voltages for transistors M3 and M5. Because transistors M3 and M5 are identical, this condition will be equivalent to their biasing at the same drain current, $I_{REF}/2$. In consequence, the aspect ratio of transistor M9 must be half the aspect ratio of transistors M7 and M8.

Table 3.1 DC operating point currents as functions of I_{REF} current

DC operating point currents	Expression
I_{D1}	$I_{REF}/2$
I_{D2}	$I_{REF}/2$
I_{D3}	$I_{REF}/2$
I_{D4}	$I_{REF}/2$
I_{D5}	$I_{REF}/2$
I_{D6}	I_{REF}
I_{D7}	I_{REF}
I_{D8}	I_{REF}
I_{D9}	$I_{REF}/2$
I_{D10}	I_{REF}

Because all biasing currents linearly depend on the reference current I_{REF} and because the current I_{REF} is strongly dependent on the supply voltage V_{DD}, on the value of the resistor R and on the values of technological parameters K'_n, K'_p and V_T, the entire current biasing of the operational amplifier having the circuit presented in Fig. 3.1 is expected to be dependent on the value of the supply voltage V_{DD} and to be affected by technologically caused errors and by temperature variations. In order to avoid this undesired behavior, the direct biasing obtained using the multiple basic current mirror M7–M10 and resistor R could be replaced by a self-biased current source (see Chap. 1).

3.2.2 *Circuit Operation and Voltage Gain*

The two stages of the operational amplifier architecture presented in Fig. 3.1 are the following:

- The first stage (between input terminals v_{IN1} and v_{IN2} and node A) is represented by an active load differential amplifier (M1–M4); transistors M1 and M2 represent the differential core, while transistors M3 and M4 are connected as a current mirror, for implementing the active load of the differential pair M1–M2.
- The second stage (between node A and node B) is implemented using transistor M5, used in common-source connection; the active load of this stage is represented by transistor M9, which is part of the multiple current mirror M7–M10.
- Transistor M6 is used as a voltage follower (between node B and output voltage v_O), for offering a small output resistance to the operational amplifier; it is used in common-drain connection, with its active load being represented by transistor M10, which is part of the current mirror M7–M10.

Using the conclusion obtained in Chap. 2 for the active load differential amplifier (Fig. 2.34, Eq. (2.89)), the absolute value of the voltage gain for the first stage (M1–M4) can be expressed as follows:

$$|a_1| = g_{m1}(r_{ds2}//r_{ds4}). \tag{3.4}$$

The absolute value of the voltage gain for the second stage (common-source connected transistor M5, having as active load transistor M9) has the following expression:

$$a_2 = -g_{m5}(r_{ds5}//r_{ds9}). \tag{3.5}$$

Because transistor M6 is used in a common-drain basic connection, it acts practically as a voltage follower, so

$$|a_{M6}| \cong 1. \tag{3.6}$$

The absolute value of the voltage gain of the entire operational amplifier presented in Fig. 3.1 will have the following expression:

$$|a| = |a_1 a_2 a_{M6}| \cong |a_1 a_2|. \tag{3.7}$$

In order to determine which is the inverting input "−" and which is the noninverting input "+" for the previously presented operational amplifier, it can be observed that between the second input voltage v_{IN2} and the output voltage v_O exist three stages: transistor M2 (common-source connection, so having a negative voltage gain), transistor M5 (also common-source connection, so having a negative voltage gain) and transistor M6 (common-drain connection, so having a positive voltage gain). In consequence, v_{IN2} will represent the noninverting input "+", and v_{IN1} will represent the inverting input "−" of the operational amplifier architecture presented in Fig. 3.1.

3.2.3 Input and Output Resistances

Because both input currents of the operational amplifier presented in Fig. 3.1 are represented by gate currents and, in consequence, they are zero, the input resistance of the circuit is infinite.

The output resistance of the operational amplifier will have the following expression (determined using Eq. (A.12) from Annex):

$$R_O \cong \frac{1}{g_{m6}}. \tag{3.8}$$

source-gate voltage of transistor M12, even they have identical aspect ratios *W*/*L* and they are biased at the same drain current. In order to determine a mathematical relation between these voltages, the reference current I_{REF} will be expressed as the drain currents of both transistors M11 and M12 (biased in saturation region):

$$I_{REF} = \frac{K'_n}{2}\left(\frac{W}{L}\right)(V_{GS11} - V_T)^2 \tag{3.17}$$

and

$$I_{REF} = \frac{K'_p}{2}\left(\frac{W}{L}\right)(V_{SG12} - V_T)^2. \tag{3.18}$$

The channel-length modulation effect is neglected in manual analysis. Computing the ratio of the previous expressions of the reference current I_{REF} results in

$$\frac{V_{GS11} - V_T}{V_{SG12} - V_T} = \sqrt{\frac{K'_p}{K'_n}}. \tag{3.19}$$

In consequence, V_{SG12} voltage can be expressed as a function of V_{GS11} voltage as follows:

$$V_{SG12} = V_T + (V_{GS11} - V_T)\sqrt{\frac{K'_n}{K'_p}}. \tag{3.20}$$

Replacing (3.20) in (3.16) and using symmetric supply voltages, $V_{SS} = -V_{DD}$, result in

$$2V_{DD} = R\frac{K'_n}{2}\left(\frac{W}{L}\right)(V_{GS11} - V_T)^2 + V_{GS11}\left(1 + \sqrt{\frac{K'_n}{K'_p}}\right) + V_T\left(1 - \sqrt{\frac{K'_n}{K'_p}}\right), \tag{3.21}$$

which represents a second-order equation, having as unknown value the gate-source voltage of transistor M11, V_{GS11}. This equation has the solutions $(V_{GS11})_A$ and $(V_{GS11})_B$, with $(V_{GS11})_A > V_T$ and $(V_{GS11})_B < V_T$. It results that $(V_{GS11})_A$ represents the mathematical solution that fulfils the condition of biasing in saturation region of transistor M11. So, replacing $(V_{GS11})_A$ in Eq. (3.17), it is possible to obtain the expression of the reference current, I_{REF}.

The expressions of the DC operating point currents of all transistors from Fig. 3.2 as a function of the reference current I_{REF} are shown in Table 3.2. All transistors are set to be identical (having the same aspect ratio *W*/*L*), except transistor M14, which has an aspect ratio half of all the other aspect ratios. The reason is the necessity of reducing the errors introduced in the operation of differential amplifier M1–M2 by the channel-length modulation effect. This can be achieved by ensuring an operation

Table 3.2 DC operating point currents as functions of I_{REF} current

DC operating point currents	Expression
I_{D1}	$I_{REF}/2$
I_{D2}	$I_{REF}/2$
I_{D3}	$I_{REF}/2$
I_{D4}	$I_{REF}/2$
I_{D5}	$I_{REF}/2$
I_{D6}	I_{REF}
I_{D7}	I_{REF}
I_{D8}	I_{REF}
I_{D9}	I_{REF}
I_{D10}	I_{REF}
I_{D11}	I_{REF}
I_{D12}	I_{REF}
I_{D13}	I_{REF}
I_{D14}	$I_{REF}/2$
I_{D15}	I_{REF}

of transistors M1 and M2 at approximately the same drain-source voltages, which is equivalent to have approximately the same gate-source voltages for transistors M3 and M5. Because transistors M3 and M5 are identical, this condition will be equivalent to their biasing at the same drain current, $I_{REF}/2$. In consequence, the aspect ratio of transistor M14 must be half the aspect ratio of transistors M12, M13 and M15.

Because all biasing currents linearly depend on the reference current I_{REF} and because I_{REF} current is strongly dependent on the supply voltage V_{DD}, on the value of resistor R and on the values of technological parameters K'_n, K'_p and V_T, the entire current biasing of the operational amplifier having the circuit presented in Fig. 3.2 is expected to be dependent on the value of the supply voltage V_{DD} and to be affected by technologically caused errors and by temperature variations. In order to avoid this undesired behavior, the direct biasing obtained using transistors M11 and M12 and resistor R could be replaced by a self-biased current source (see Chap. 1).

3.3.2 Circuit Operation and Voltage Gain

The three stages of the operational amplifier architecture presented in Fig. 3.2 are the following:

- The first stage (between input terminals v_{IN1} and v_{IN2} and node A) is represented by an active load differential amplifier (M1–M4); transistors M1 and M2 represent the differential core, while transistors M3 and M4 are connected as a current mirror, for implementing the active load of the differential core M1–M2.

- The second stage (between node A and node B) is implemented using transistor M5, used in common-source connection; the active load of this stage is represented by transistor M14, which is part of multiple current mirror M12–M15.
- The third stage (between node B and node C) is realized using transistor M6, also used in common-source connection; the output current of transistor M6 is mirrored using the current mirror M7–M8, with the drain current of transistor M8 having as active load M15 transistor, which is also part of multiple current mirror M12–M15.
- Transistor M9 is used as a voltage follower (between node C and output voltage v_O), for offering a small output resistance to the operational amplifier; it is used in common-drain connection, with its active load being represented by transistor M10, which is part of M10–M11 current mirror.

Using the conclusion obtained in Chap. 2 for the active load differential amplifier (Fig. 2.34, Eq. (2.89)), the absolute value of the voltage gain for the first stage (M1–M4) can be expressed as follows:

$$|a_1| = g_{m1}(r_{ds2}//r_{ds4}). \tag{3.22}$$

The absolute value of the voltage gain for the second stage (common-source connected transistor M5, having as active load transistor M14) has the following expression:

$$|a_2| = g_{m5}(r_{ds5}//r_{ds14}). \tag{3.23}$$

The absolute value of the voltage gain for the third stage (common-source connected transistor M6, having as active load transistor M15, by using the current mirror M7–M8) can be expressed as follows:

$$|a_3| = g_{m6}\frac{(W/L)_8}{(W/L)_7}(r_{ds8}//r_{ds15}), \tag{3.24}$$

because the transfer ratio of the current mirror M7–M8 is equal to $\frac{(W/L)_8}{(W/L)_7}$.

As transistor M9 is used in a common-drain basic connection, it is practically a voltage follower, so

$$|a_{M9}| \cong 1. \tag{3.25}$$

The absolute value of the voltage gain of the entire operational amplifier presented in Fig. 3.2 will have the following expression:

$$|a| = |a_1 a_2 a_3 a_{M9}| \cong |a_1 a_2 a_3|. \tag{3.26}$$

In order to determine which is the inverting input "−" and which is the noninverting input "+" for the previously presented operational amplifier, the following analysis can be performed:

$$v_{IN2} \uparrow v_{sg2} \downarrow i_{d2} \downarrow v_A \downarrow v_{gs5} \downarrow i_{d5} \downarrow . \tag{3.27}$$

In consequence,

$$v_B \uparrow v_{sg6} \downarrow i_{d6} \downarrow i_{d8} \downarrow v_C \uparrow . \tag{3.28}$$

Because M19 is a voltage follower, it results that $v_O\uparrow$. In conclusion, v_{IN2} represents the noninverting input "+" and, in consequence, v_{IN1} represents the inverting input "−".

3.3.3 Input and Output Resistances

Because both input currents of the operational amplifier presented in Fig. 3.2 are represented by gate currents and, in consequence, they are zero, the input resistance of the circuit is infinite.

The output resistance of the operational amplifier will have the following expression (determined using Eq. (A.12) from Annex):

$$R_O \cong \frac{1}{g_{m9}}. \tag{3.29}$$

3.3.4 Maximal Range of the Common-Mode Input Voltage

In order to determine the maximal value of the common-mode input voltage for the operational amplifier having the architecture presented in Fig. 3.2, suppose that the common-mode input voltage $v_{IN1} = v_{IN2} \overset{not}{=} V_{IC}$ is increasing. Because the source-gate voltages of transistors M1 and M2 are constant, the drain potential of transistor M13 will also increase. The maximal value of this potential is given by the condition that transistor M13 be still biased in saturation region. In consequence, the maximal value of the common-mode input voltage will have the following expression:

$$V_{IC}^{\mathrm{MAX}} = V_{DD} - V_{SD13sat} - V_{SG1}, \tag{3.30}$$

equivalent to

$$V_{IC}^{\mathrm{MAX}} = V_{DD} - V_T - \left(\sqrt{2}+1\right)\sqrt{\frac{I_{REF}}{K'_p\left(\frac{W}{L}\right)}}. \tag{3.31}$$

In order to determine the minimal value of the common-mode input voltage for the operational amplifier having the architecture presented in Fig. 3.2, suppose that the common-mode input voltage $v_{IN1} = v_{IN2} \overset{not}{=} V_{IC}$ is decreasing. Because the gate-source voltages of transistors M3 and M5 are constant (having the same value), the minimal value of the common-mode input voltage will be given by the condition that transistors M1 and M2 be still biased in saturation region. In consequence, the minimal value of the common-mode input voltage will have the following expression:

$$V_{IC}^{\mathrm{MIN}} = V_{SS} + V_{GS3} + V_{SD1sat} - V_{SG1}, \tag{3.32}$$

equivalent to

$$V_{IC}^{\mathrm{MIN}} = V_{SS} + V_{GS3} - V_{T}, \tag{3.33}$$

or

$$V_{IC}^{\mathrm{MIN}} = V_{SS} + \sqrt{\frac{I_{REF}}{K'_{n}\left(\frac{W}{L}\right)}}. \tag{3.34}$$

3.4 Two-Stage CMOS Cascode Operational Amplifier (Architecture 3 of a CMOS Operational Amplifier)

An alternative approach of designing an architecture of an operational amplifier is to use a cascode configuration, in order to strongly increase the differential-mode voltage gain of the circuit [1–4]. The disadvantage of this class of architectures of operational amplifiers is mainly related to the increasing of the minimal supply voltage, limiting the utilization of these circuits to the applications that have a relatively large value of the supply voltage available. In Fig. 3.3, a two-stage CMOS cascode operational amplifier is presented.

In order to evaluate the performance of the two-stage CMOS cascode operational amplifier, the following parameters must be evaluated.

3.4.1 DC Operating Point Currents

Noting with I_{REF} the current, which flows through resistor R, all biasing currents of the operational amplifier will linearly depend on the reference current I_{REF}. In order to compute this current, the following equation can be written as

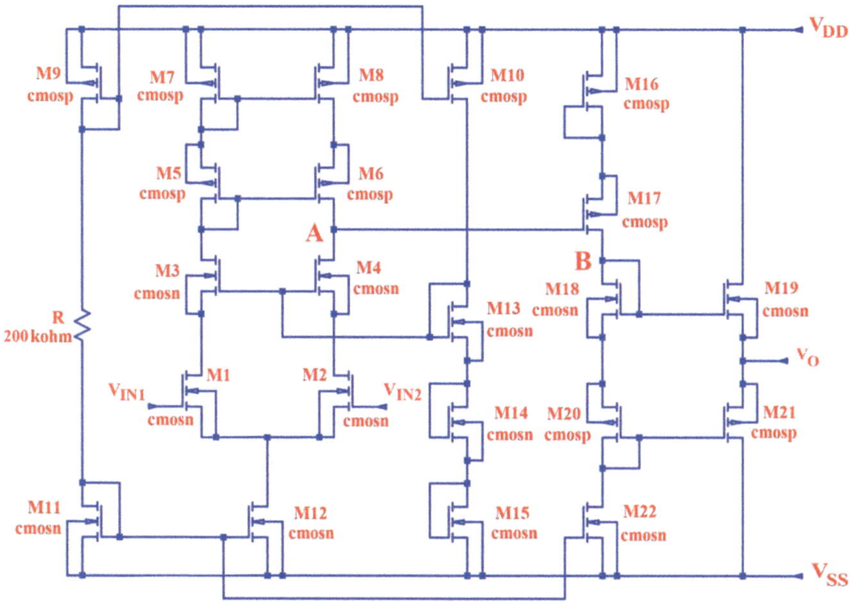

Fig. 3.3 Two-stage CMOS cascode operational amplifier architecture

$$V_{DD} - V_{SS} = RI_{REF} + V_{GSsy} + V_{SG9}. \tag{3.35}$$

Because the performances of NMOS and PMOS transistors are different ($K'_n \neq K'_p$), the gate-source voltage of transistor M11 will be different to the source-gate voltage of transistor M9, even when they have identical aspect ratios W/L and they are biased at the same drain current. In order to determine a mathematical relation between these voltages, the reference current I_{REF} will be expressed as the drain currents of both transistors M11 and M12 (biased in saturation region):

$$I_{REF} = \frac{K'_n}{2}\left(\frac{W}{L}\right)(V_{GS11} - V_T)^2 \tag{3.36}$$

and

$$I_{REF} = \frac{K'_p}{2}\left(\frac{W}{L}\right)(V_{SG9} - V_T)^2. \tag{3.37}$$

The channel-length modulation effect is neglected in manual analysis. Computing the ratio of the previous expressions of the reference current I_{REF} results in

$$\frac{V_{GS11} - V_T}{V_{SG9} - V_T} = \sqrt{\frac{K'_p}{K'_n}}. \tag{3.38}$$

In consequence, V_{SG12} voltage can be expressed as a function of V_{GS11} voltage as follows:

$$V_{SG12} = V_T + (V_{GS11} - V_T)\sqrt{\frac{K'_n}{K'_p}}. \tag{3.39}$$

Replacing (3.39) in (3.35) and using symmetric supply voltages, $V_{SS} = -V_{DD}$, result in

$$2V_{DD} = R\frac{K'_n}{2}\left(\frac{W}{L}\right)(V_{GS11} - V_T)^2 + V_{GS11}\left(1 + \sqrt{\frac{K'_n}{K'_p}}\right) + V_T\left(1 - \sqrt{\frac{K'_n}{K'_p}}\right), \tag{3.40}$$

which represents a second-order equation, having as unknown value the gate-source voltage of transistor M11, V_{GS11}. This equation has the solutions $(V_{GS11})_A$ and $(V_{GS11})_B$, with $(V_{GS11})_A > V_T$ and $(V_{GS11})_B < V_T$. It results that $(V_{GS11})_A$ represents the mathematical solution that fulfils the condition of biasing in saturation region of transistor M11. So, replacing $(V_{GS11})_A$ in Eq. (3.36), it is possible to obtain the expression of the reference current, I_{REF}. All transistors are set to be identical (having the aspect ratio (W/L)), except transistors M13, M14 and M15, which have aspect ratios equal to $(W/L)/10$ and transistor M22, which has an aspect ratio equal to $(W/L)/2$.

In order to compute the DC operating point current of final transistors M19 and M21, noted with I in the following relations, the translinear loop containing transistors M18, M19, M20 and M21 must be analysed:

$$V_{GS18} + V_{SG20} = V_{GS19} + V_{SG21}, \tag{3.41}$$

equivalent to

$$\left[\sqrt{\frac{I_{REF}}{K'_n\left(\frac{W}{L}\right)}} + V_T\right] + \left[\sqrt{\frac{I_{REF}}{K'_p\left(\frac{W}{L}\right)}} + V_T\right] = \left[\sqrt{\frac{2I}{K'_n\left(\frac{W}{L}\right)}} + V_T\right] + \left[\sqrt{\frac{2I}{K'_p\left(\frac{W}{L}\right)}} + V_T\right], \tag{3.42}$$

resulting in $I = I_{REF}/2$.

The expressions of DC operating point currents of all transistors from Fig. 3.3 as a function of the reference current I_{REF} are shown in Table 3.3.

Because all biasing currents linearly depend on the reference current I_{REF} and because the current I_{REF} is strongly dependent on the supply voltage V_{DD}, on the value of resistor R and on the values of technological parameters K'_n, K'_p and V_T, the entire current biasing of the operational amplifier having the circuit presented in Fig. 3.3 is expected to be dependent on the value of the supply voltage V_{DD} and to be affected by technologically caused errors and by temperature variations. In order to

Table 3.3 DC operating point currents as functions of I_{REF} current

DC operating point currents	Expression
I_{D1}	$I_{REF}/2$
I_{D2}	$I_{REF}/2$
I_{D3}	$I_{REF}/2$
I_{D4}	$I_{REF}/2$
I_{D5}	$I_{REF}/2$
I_{D6}	$I_{REF}/2$
I_{D7}	$I_{REF}/2$
I_{D8}	$I_{REF}/2$
I_{D9}	I_{REF}
I_{D10}	I_{REF}
I_{D11}	I_{REF}
I_{D12}	I_{REF}
I_{D13}	I_{REF}
I_{D14}	I_{REF}
I_{D15}	I_{REF}
I_{D16}	$I_{REF}/2$
I_{D17}	$I_{REF}/2$
I_{D18}	$I_{REF}/2$
I_{D19}	$I_{REF}/2$
I_{D20}	$I_{REF}/2$
I_{D21}	$I_{REF}/2$
I_{D22}	$I_{REF}/2$

avoid this undesired behavior, the direct biasing obtained using transistors M11 and M9 and resistor R could be replaced by a self-biased current source (see Chap. 1).

3.4.2 Circuit Operation and Voltage Gain

The two stages of the operational amplifier architecture presented in Fig. 3.3 are the following:

- The first stage (between input terminals v_{IN1} and v_{IN2} and node A) is represented by an active load differential amplifier (M1–M8); transistors M1 and M2 represent the differential core, while transistors M5–M8 are connected as a cascode current mirror, for implementing the active load of the differential amplifier M1–M2; the pair of transistors M3 and M4 is used for obtaining an extremely large resistance of node A, required for strongly increasing the voltage gain of this stage.
- The second stage (between node A and node B) is implemented using transistor M17, operating as a source-degenerated stage, where the source impedance is implemented using a gate-drain-connected MOS transistor M16; the impedance of transistor M16 introduces local feedback, modifying both the effective

transconductance and the output resistance of the amplifier; the load of this stage is practically represented by transistor M22, which is part of multiple current mirror M11–M12–M22.

- The output stage (between node B and the output voltage v_O) is realized using transistors M18, M19, M20 and M21.

Extrapolating the conclusion obtained in Chap. 2 for the active load differential amplifier (Fig. 2.34, Eq. (2.89)), the absolute value of the voltage gain for the first stage (M1–M8) can be expressed as follows (using Eq. (A.10) from Annex):

$$|a_1| = g_{m1}\left[\left(g_{m2}r_{ds2}^2\right)//\left(g_{m6}r_{ds6}^2\right)\right], \tag{3.43}$$

where the output resistance of the classical stage is replaced with the output resistance of the new cascode stage. The utilization of the cascode stage strongly increases the voltage gain of the first stage of the operational amplifier presented in Fig. 3.3.

The absolute value of the voltage gain for the second stage can be expressed as follows:

$$|a_2| = \frac{g_{m17}R_{load}}{1 + g_{m17}R_S}, \tag{3.44}$$

where R_{load} and R_S represent the equivalent load resistance and the equivalent resistance connected in the source terminal of transistor M17, respectively.

The equivalent resistance connected in the source terminal of transistor M17, R_S, is represented by the gate-drain-connected transistor M16, having an equivalent resistance equal to $1/g_{m16}$.

As the equivalent resistances of transistors M18 and M20 are extremely small ($1/g_{m18}$ and $1/g_{m20}$, respectively) in comparison with the equivalent resistance view in the drain of transistor M22 (r_{ds22}), it results that the load resistance of this stage R_{load} will have the following expression (determined using Eq. (A.10) from Annex):

$$R_{load} = r_{ds22}//r_{ds17}\left[1 + g_{m17}(1/g_{m16})\right]. \tag{3.45}$$

Because transistors M16 and M17 are identical and they are biased at the same current $I_{REF}/2$, it results that $g_{m16} = g_{m17}$. So, the previous relation becomes

$$R_{load} = r_{ds22}//2r_{ds17}. \tag{3.46}$$

In conclusion, the expression of the absolute value of the voltage gain for the second stage will be

$$|a_2| = \frac{g_{m17}}{2}\left(r_{ds22}//2r_{ds17}\right). \tag{3.47}$$

As both transistors M19 and M21 are used in a basic common-drain connection, each represents practically a voltage follower, so

$$|a_{M9-M21}| \cong 1. \tag{3.48}$$

The absolute value of the voltage gain of the operational amplifier presented in Fig. 3.3 will be $|a| = |a_1 a_2 a_{M9-M21}| \cong |a_1 a_2|$. Because of the utilization of a cascode architecture for implementing the first amplifier stage, this stage will have an extremely large value of the voltage gain, much greater than the voltage gain of the second amplifier stage.

In order to determine which is the inverting input "–" and which is the noninverting input "+" for the previously presented operational amplifier, it can be observed that between the second input voltage v_{IN2} and the output voltage v_O exist three stages: transistor M2 (common-source connection, so having a negative voltage gain), transistor M17 (source-degenerated connection, so having a negative voltage gain) and transistor M19 or M21 (common-drain connection, so having a positive voltage gain). In consequence, v_{IN2} will represent the noninverting input "+", and v_{IN1} will represent the inverting input "–" of the operational amplifier architecture presented in Fig. 3.3.

3.4.3 Input and Output Resistances

Because both input currents of the operational amplifier presented in Fig. 3.3 are represented by gate currents and, in consequence, they are zero, the input resistance of the circuit is infinite.

The output resistance of the operational amplifier will have the following expression:

$$R_O \cong \frac{1}{g_{m19} + g_{m21}}. \tag{3.49}$$

3.4.4 Maximal Range of the Common-Mode Input Voltage

In order to determine the maximal value of the common-mode input voltage for the operational amplifier having the architecture presented in Fig. 3.3, suppose that the common-mode input voltage $v_{IN1} = v_{IN2} \overset{not}{=} V_{IC}$ is increasing. As the gate-source voltages of transistors M1 and M2 are constant, the consequence will be an increasing of the potential from the drain of transistor M12. Because the drain potentials of transistors M1 and M2 are constant and equal to

$$V_{D(M1)} = V_{D(M2)} = V_{SS} + V_{GS13} + V_{GS14} + V_{GS15} - V_{GS4}, \tag{3.50}$$

the maximal value of the common-mode input voltage will be given by the condition that transistors M1 and M2 be still biased in saturation region. In consequence, the maximal value of the common-mode input voltage will have the following expression:

$$V_{IC}^{\text{MAX}} = V_{D(M1)} - V_{DS1sat} + V_{GS1}, \tag{3.51}$$

equivalent to

$$V_{IC}^{\text{MAX}} = V_{D(M1)} + V_T. \tag{3.52}$$

Evaluating the expression of potential $V_{D(M1)}$ using Eq. (3.50) and replacing it in Eq. (3.52) result in

$$V_{IC}^{\text{MAX}} = V_{SS} + 3V_T + \left(6\sqrt{5} - 1\right)\sqrt{\frac{I_{REF}}{K'_n\left(\frac{W}{L}\right)}}. \tag{3.53}$$

In order to determine the minimal value of the common-mode input voltage for the operational amplifier having the architecture presented in Fig. 3.3, suppose that the common-mode input voltage $v_{IN1} = v_{IN2} \overset{not}{=} V_{IC}$ is decreasing.

Because gate-source voltages of transistors M1 and M2 are constant, the drain potential of transistor M12 will also decrease. The minimal value of this potential is given by the condition that transistor M12 be still biased in saturation region. In consequence, the minimal value of the common-mode input voltage will have the following expression:

$$V_{IC}^{\text{MIN}} = V_{SS} + V_{DS12sat} + V_{GS1}, \tag{3.54}$$

equivalent to

$$V_{IC}^{\text{MIN}} = V_{SS} + (V_{GS12} - V_T) + V_{GS1}, \tag{3.55}$$

or

$$V_{IC}^{\text{MIN}} = V_{SS} + V_T + \left(\sqrt{2} + 1\right)\sqrt{\frac{I_{REF}}{K'_n\left(\frac{W}{L}\right)}}. \tag{3.56}$$

3.5 One-Stage NMOS Folded Cascode Operational Amplifier (Architecture 4 of a CMOS Operational Amplifier)

The folded cascode operational amplifier is a classical high-gain CMOS architecture, widely used in analog integrated circuit design and extensively described in literature [1–4]. It has the important advantage of strongly increasing the voltage gain, limiting also the increasing of the minimal supply voltage in comparison with the cascode operational amplifier architectures. In Fig. 3.4, a one-stage NMOS folded cascode operational amplifier is presented.

In order to evaluate the performance of the one-stage NMOS folded cascode operational amplifier, the following parameters must be evaluated.

3.5.1 DC Operating Point Currents

Noting with I_{REF} the current passing through resistor R, all biasing currents of the operational amplifier will linearly depend on the reference current I_{REF}. This reference current is made by using a self-biased current source, implemented using transistors M4–M7 and resistor R (see Chap. 1). In order to compute this current, the following equation can be written:

$$V_{GS4} = V_{GS5} + RI_{REF}. \tag{3.57}$$

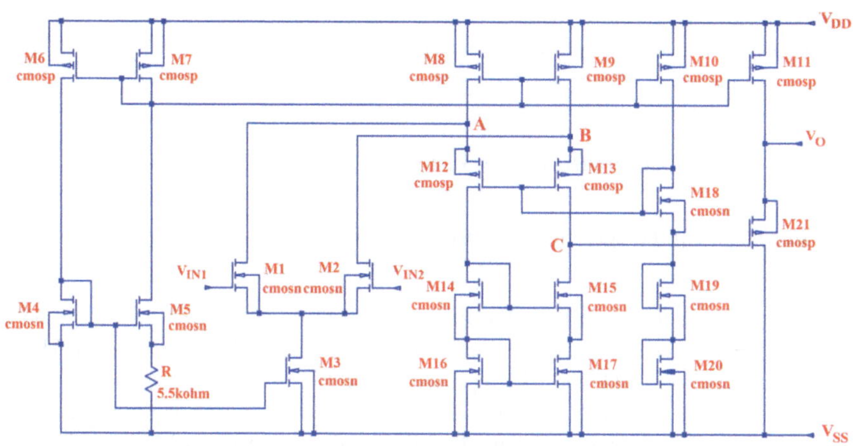

Fig. 3.4 One-stage NMOS folded cascode operational amplifier architecture

Choosing, for example, an aspect ratio of the transistor M5 four times greater than the aspect ratio of transistor M4 (in order to introduce a controllable asymmetry for a proper operation of the circuit) results in

$$V_T + \sqrt{\frac{2I_{REF}}{K'_n\left(\frac{W}{L}\right)}} = V_T + \sqrt{\frac{2I_{REF}}{4K'_n\left(\frac{W}{L}\right)}} + RI_{REF}. \tag{3.58}$$

So, the expression of the reference current I_{REF} will be

$$I_{REF} = \frac{1}{2R^2K'_n\left(\frac{W}{L}\right)}. \tag{3.59}$$

The resulted expression of the reference current I_{REF} does not depend on the supply voltage variations because of the utilization of the self-biasing method and neglecting of the channel-length modulation effect. The additional consideration in computations of this undesired effect will introduce a small dependence $I_{REF}(V_{DD})$, much inferior to the value obtained for the current sources without self-biasing techniques (see Chap. 1).

The expressions of DC operating point currents of all transistors from Fig. 3.4 as a function of the reference current I_{REF} are shown in Table 3.4. All transistors are set to be identical (having W/L parameter), except transistor M5, which has an aspect ratio equal to $4(W/L)$, and transistor M10, which has an aspect ratio equal to $3(W/L)$.

Because all biasing currents linearly depend on the reference current I_{REF} and because I_{REF} current does not dependent on the supply voltage V_{DD}, the entire current biasing of the operational amplifier having the circuit presented in Fig. 3.4 is expected to be independent (in a first-order analysis) on the value of the supply voltage V_{DD}. On the other hand, the same biasing is affected by technologically caused errors and by temperature variations.

3.5.2 Circuit Operation and Voltage Gain

The two stages of the operational amplifier architecture presented in Fig. 3.4 are the following:

- The first stage (between input terminals v_{IN1} and v_{IN2} and node C) is represented by a folded cascode differential amplifier (transistors M1, M2, M8, M9 and M12–M17); transistors M1 and M2 represent the differential core, while transistors M14–M17 are connected as a cascode current mirror, for implementing the active load of the differential amplifier M1–M2; a pair of transistors M12 and M13 is used for obtaining an extremely large resistance of node C, required for strongly increasing the voltage gain of this stage.

Table 3.4 DC operating point currents as functions of I_{REF} current

DC operating point currents	Expression
I_{D1}	$I_{REF}/2$
I_{D2}	$I_{REF}/2$
I_{D3}	I_{REF}
I_{D4}	I_{REF}
I_{D5}	I_{REF}
I_{D6}	I_{REF}
I_{D7}	I_{REF}
I_{D8}	I_{REF}
I_{D9}	I_{REF}
I_{D10}	$3I_{REF}$
I_{D11}	I_{REF}
I_{D12}	$I_{REF}/2$
I_{D13}	$I_{REF}/2$
I_{D14}	$I_{REF}/2$
I_{D15}	$I_{REF}/2$
I_{D16}	$I_{REF}/2$
I_{D17}	$I_{REF}/2$
I_{D18}	$3I_{REF}$
I_{D19}	$3I_{REF}$
I_{D20}	$3I_{REF}$
I_{D21}	I_{REF}

- Transistor M21 is used as a voltage follower (between node C and output voltage v_O), offering a small output resistance to the operational amplifier; it is used in common-drain connection, with its active load being represented by transistor M11, which is part of M6–M11 multiple current mirror.

In order to determine the expression of the voltage gain for the first folded cascode differential amplifier stage, the differential output current in node C must be computed. This dynamic current, noted with i_{O1}, can be expressed as follows:

$$i_{O1} = i_{D13} - i_{D15} = (I_{D9} - i_{D2}) - (I_{D8} - i_{D1}) = i_{D1} - i_{D2}, \tag{3.60}$$

because the currents I_{D8} and I_{D9} represent the output currents of a current mirror and they are equal, $I_{D8} = I_{D9} = I_{REF}$. So, all the previously folded cascode differential amplifier (transistors M1, M2, M8, M9 and M12–M17) is functionally equivalent to the classical active load differential amplifier from Fig. 2.34. In consequence, the absolute value of its voltage gain can be expressed, similarly to Eq. (2.89), as follows:

$$|a_1| = g_{m1} R_{O1}, \tag{3.61}$$

where R_{O1} represents the output impedance of the first amplifier stage in node C (determined using Eq. (A.10) from Annex):

$$R_{O1} = r_{ds15}(1 + g_{m15}r_{ds17})//r_{ds13}[1 + g_{m13}(r_{ds2}//r_{ds9})], \tag{3.62}$$

approximately equal to

$$R_{O1} = g_{m15}r_{ds15}^2//g_{m13}r_{ds13}(r_{ds2}//r_{ds9}), \tag{3.63}$$

resulting in the following expression of the absolute value of the voltage gain a_1:

$$|a_1| = g_{m1}\left[g_{m15}r_{ds15}^2//g_{m13}r_{ds13}(r_{ds2}//r_{ds9})\right]. \tag{3.64}$$

Because transistor M21 is used in a common-drain basic connection, it represents practically a voltage follower, so

$$|a_{M21}| \cong 1. \tag{3.65}$$

The absolute value of the voltage gain of the operational amplifier presented in Fig. 3.4 will be $|a| = |a_1 a_{M21}| \cong |a_1|$.

In order to determine which is the inverting input "−" and which is the noninverting input "+" for the previously presented operational amplifier, the following analysis can be performed:

$$v_{IN2} \uparrow v_{gs2} \uparrow i_{d2} \uparrow i_{d13} \downarrow v_C \downarrow. \tag{3.66}$$

Because M21 is a voltage follower, it results that $v_O\downarrow$. In conclusion, v_{IN2} represents the inverting input "−" and, in consequence, v_{IN1} represents the noninverting input "+".

3.5.3 *Input and Output Resistances*

Because both input currents of the operational amplifier presented in Fig. 3.4 are represented by gate currents and, in consequence, they are zero, the input resistance of the circuit is infinite.

The output resistance of the operational amplifier will have the following expression (determined using Eq. (A.12) from Annex):

$$R_O \cong \frac{1}{g_{m21}}. \tag{3.67}$$

3.5.4 Maximal Range of the Common-Mode Input Voltage

In order to determine the maximal value of the common-mode input voltage for the operational amplifier having the architecture presented in Fig. 3.4, suppose that the common-mode input voltage $v_{IN1} = v_{IN2} \stackrel{not}{=} V_{IC}$ is increasing. As the gate-source voltages of transistors M1 and M2 are constant, the consequence will be an increasing of the potential from the drain of transistor M3. Because the drain potentials of transistors M1 and M2 are constant and equal to

$$V_{D(M1)} = V_{D(M2)} = V_{SS} + V_{GS18} + V_{GS19} + V_{GS20} + V_{SG12}, \tag{3.68}$$

the maximal value of the common-mode input voltage will be given by the condition that transistors M1 and M2 be still biased in saturation region. In consequence, the maximal value of the common-mode input voltage will have the following expression:

$$V_{IC}^{\mathrm{MAX}} = V_{D(M1)} - V_{DS1sat} + V_{GS1}, \tag{3.69}$$

equivalent to

$$V_{IC}^{\mathrm{MAX}} = V_{D(M1)} + V_T. \tag{3.70}$$

Evaluating the expression of potential $V_{D(M1)}$ using Eq. (3.68) and replacing it in Eq. (3.70) result in

$$V_{IC}^{\mathrm{MAX}} = V_{SS} + 5V_T + 3\sqrt{\frac{6I_{REF}}{K'_n\left(\frac{W}{L}\right)}} + \sqrt{\frac{I_{REF}}{K'_p\left(\frac{W}{L}\right)}}. \tag{3.71}$$

In order to determine the minimal value of the common-mode input voltage for the operational amplifier having the architecture presented in Fig. 3.4, suppose that the common-mode input voltage $v_{IN1} = v_{IN2} \stackrel{not}{=} V_{IC}$ is decreasing. Because the gate-source voltages of transistors M1 and M2 are constant, the drain potential of transistor M3 will also decrease. The minimal value of this potential is given by the condition that transistor M3 be still biased in saturation region. In consequence, the minimal value of the common-mode input voltage will have the following expression:

$$V_{IC}^{\mathrm{MIN}} = V_{SS} + V_{DS3sat} + V_{GS1}, \tag{3.72}$$

equivalent to

$$V_{IC}^{\text{MIN}} = V_{SS} + (V_{GS3} - V_T) + V_{GS1}, \tag{3.73}$$

or

$$V_{IC}^{\text{MIN}} = V_{SS} + V_T + \left(\sqrt{2} + 1\right)\sqrt{\frac{I_{REF}}{K'_n\left(\frac{W}{L}\right)}}. \tag{3.74}$$

3.6 One-Stage PMOS Folded Cascode Operational Amplifier (Architecture 5 of a CMOS Operational Amplifier)

Folded cascode topologies represent a standard extension of the cascode technique, enabling higher-voltage headroom in low-supply CMOS technologies [1–4]. The NMOS input core from Fig. 3.4 is replaced by a complementary PMOS input core in Fig. 3.5 (transistors M1 and M2 for both circuits). The fundamental difference between these two similar architectures is mainly related to the maximal range of the common-mode input voltage. Usually, an NMOS differential amplifier works excellent for large values of the common-mode input voltages and does not work for very small values of the common-mode input voltages. Complementary, a PMOS differential amplifier works excellent for small values of the common-mode input voltages and does not work for very large values of the common-mode input voltages. The decision for choosing one or other architecture is mainly influenced by the application in which the operational amplifier will be used.

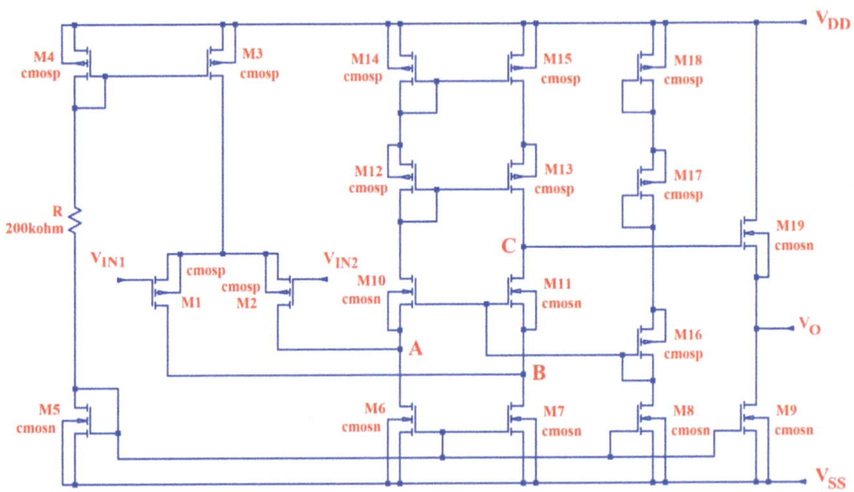

Fig. 3.5 One-stage PMOS folded cascode operational amplifier architecture

In order to evaluate the performance of the one-stage PMOS folded cascode operational amplifier, the following parameters must be evaluated.

3.6.1 DC Operating Point Currents

Noting with I_{REF} the current, which flows through the resistor R, all biasing currents of the operational amplifier will linearly depend on the reference current I_{REF}. In order to compute this current, the following equation can be written:

$$V_{DD} - V_{SS} = RI_{REF} + V_{SG4} + V_{GS5}. \tag{3.75}$$

Because the performances of NMOS and PMOS transistors are different ($K'_n \neq K'_p$), the source-gate voltage of transistor M4 will be different to the gate-source voltage of transistor M5, even if they have identical aspect ratios W/L and they are biased at the same drain current. In order to determine a mathematical relation between these voltages, the reference current I_{REF} will be expressed as the drain currents of both transistors M4 and M5 (biased in saturation region):

$$I_{REF} = \frac{K'_n}{2}\left(\frac{W}{L}\right)(V_{GS5} - V_T)^2 \tag{3.76}$$

and

$$I_{REF} = \frac{K'_p}{2}\left(\frac{W}{L}\right)(V_{SG4} - V_T)^2. \tag{3.77}$$

The channel-length modulation effect is neglected in manual analysis. Computing the ratio of the previous expressions of the reference current I_{REF} results in

$$\frac{V_{GS5} - V_T}{V_{SG4} - V_T} = \sqrt{\frac{K'_p}{K'_n}}. \tag{3.78}$$

In consequence, V_{SG4} voltage can be expressed as a function of V_{GS5} voltage as follows:

$$V_{SG4} = V_T + (V_{GS5} - V_T)\sqrt{\frac{K'_n}{K'_p}}. \tag{3.79}$$

Replacing Eq. (3.79) in Eq. (3.75) and using symmetric supply voltages, $V_{SS} = -V_{DD}$, result in

$$2V_{DD} = R\frac{K'_n}{2}\left(\frac{W}{L}\right)(V_{GS5} - V_T)^2 + V_{GS5}\left(1 + \sqrt{\frac{K'_n}{K'_p}}\right) + V_T\left(1 - \sqrt{\frac{K'_n}{K'_p}}\right), \tag{3.80}$$

Table 3.5 DC operating point currents as functions of I_{REF} current

DC operating point currents	Expression
I_{D1}	$I_{REF}/2$
I_{D2}	$I_{REF}/2$
I_{D3}	I_{REF}
I_{D4}	I_{REF}
I_{D5}	I_{REF}
I_{D6}	I_{REF}
I_{D7}	I_{REF}
I_{D8}	$3I_{REF}$
I_{D9}	I_{REF}
I_{D10}	$I_{REF}/2$
I_{D11}	$I_{REF}/2$
I_{D12}	$I_{REF}/2$
I_{D13}	$I_{REF}/2$
I_{D14}	$I_{REF}/2$
I_{D15}	$I_{REF}/2$
I_{D16}	$3I_{REF}$
I_{D17}	$3I_{REF}$
I_{D18}	$3I_{REF}$
I_{D19}	I_{REF}

which represents a second-order equation, having as unknown value the gate-source voltage of transistor M5, V_{GS5}. This equation has the solutions $(V_{GS5})_A$ and $(V_{GS5})_B$, with $(V_{GS5})_A > V_T$ and $(V_{GS5})_B < V_T$. It results that $(V_{GS5})_A$ represents the mathematical solution that fulfils the condition of biasing in saturation region of transistor M5. So, replacing $(V_{GS5})_A$ in Eq. (3.76), it is possible to obtain the expression of the reference current, I_{REF}.

The expressions of DC operating point currents of all transistors from Fig. 3.5 as a function of the reference current I_{REF} are shown in Table 3.5. All transistors are set identical (having W/L parameter), except transistor M8, which has an aspect ratio equal to $3(W/L)$.

Because all biasing currents linearly depend on the reference current I_{REF} and because I_{REF} current is strongly dependent on the supply voltage V_{DD}, on the value of the resistor R and on the values of technological parameters K'_n, K'_p and V_T, the entire current biasing of the operational amplifier having the circuit presented in Fig. 3.5 is expected to be dependent on the value of the supply voltage V_{DD} and to be affected by technologically caused errors and by temperature variations. In order to avoid this undesired behavior, the direct biasing obtained using transistors M4 and M5 and resistor R could be replaced by a self-biased current source (see Chap. 1).

3.6.2 Circuit Operation and Voltage Gain

The amplifier stages of the operational amplifier architecture presented in Fig. 3.5 are the following:

- The first stage (between input terminals v_{IN1} and v_{IN2} and node C) is represented by a folded cascode differential amplifier (transistors M1, M2, M6, M7 and M10–M15); transistors M1 and M2 represent the differential core, while transistors M12–M15 are connected as a cascode current mirror, for implementing the active load of the differential amplifier M1–M2; a pair of transistors M10 and M11 is used for obtaining an extremely large resistance of node C, required for strongly increasing the voltage gain of this stage.
- Transistor M19 is used as a voltage follower (between node C and output voltage v_O), offering a small output resistance to the operational amplifier; it is used in common-drain connection, with its active load being represented by transistor M9, which is part of M5–M9 multiple current mirror.

In order to determine the expression of the voltage gain for the first folded cascode differential amplifier stage, the differential output current in node C must be computed. This dynamic current, noted with i_{O1}, can be expressed as follows:

$$i_{O1} = i_{D13} - i_{D11} = (I_{D6} - i_{D2}) - (I_{D7} - i_{D1}) = i_{D1} - i_{D2}, \tag{3.81}$$

because the currents I_{D6} and I_{D7} represent the output currents of a current mirror and they are equal, $I_{D6} = I_{D7} = I_{REF}$. So, all the previous folded cascode differential amplifier (transistors M1, M2, M6, M7 and M10–M15) is functionally equivalent to the classical active load differential amplifier from Fig. 2.34. In consequence, the absolute value of its voltage gain can be expressed, similarly to Eq. (2.89), as follows:

$$|a_1| = g_{m1} R_{O1}, \tag{3.82}$$

where R_{O1} represents the output impedance of the first amplifier stage (determined using Eq. (A.10) from Annex):

$$R_{O1} = r_{ds13}(1 + g_{m13} r_{ds15}) // r_{ds11}[1 + g_{m11}(r_{ds1} // r_{ds7})], \tag{3.83}$$

approximately equal to

$$R_{O1} = g_{m13} r_{ds13}^2 // g_{m11} r_{ds11}(r_{ds1} // r_{ds7}), \tag{3.84}$$

resulting in the following expression of the absolute value of the voltage gain a_1:

$$|a_1| = g_{m1}\left[g_{m13} r_{ds13}^2 // g_{m11} r_{ds11}(r_{ds1} // r_{ds7})\right]. \tag{3.85}$$

Because transistor M19 is used in a common-drain basic connection, it represents practically a voltage follower, so

$$|a_{M19}| \cong 1. \tag{3.86}$$

The absolute value of the voltage gain of the operational amplifier presented in Fig. 3.5 will be $|a| = |a_1 a_{M19}| \cong |a_1|$.

In order to determine which is the inverting input "−" and which is the noninverting input "+" for the previously presented operational amplifier, the following analysis can be performed:

$$v_{IN1} \uparrow v_{sg1} \downarrow i_{d1} \downarrow i_{d11} \uparrow v_C \downarrow . \tag{3.87}$$

Because M19 is a voltage follower, it results that $v_O\downarrow$. In conclusion, v_{IN1} represents the inverting input "−" and, in consequence, v_{IN2} represents the noninverting input "+".

3.6.3 *Input and Output Resistances*

Because both input currents of the operational amplifier presented in Fig. 3.5 are represented by gate currents and, in consequence, they are zero, the input resistance of the circuit is infinite.

The output resistance of the operational amplifier will have the following expression (determined using Eq. (A.12) from Annex):

$$R_O \cong \frac{1}{g_{m19}}. \tag{3.88}$$

3.6.4 *Maximal Range of the Common-Mode Input Voltage*

In order to determine the minimal value of the common-mode input voltage for the operational amplifier having the architecture presented in Fig. 3.5, suppose that the common-mode input voltage $v_{IN1} = v_{IN2} \overset{not}{=} V_{IC}$ is decreasing. As the source-gate voltages of transistors M1 and M2 are constant, the consequence will be a decreasing of the potential from the drain of transistor M3. Because the drain potentials of transistors M1 and M2 are constant and equal to

$$V_{D(M1)} = V_{D(M2)} = V_{DD} - V_{SG18} - V_{SG17} - V_{SG16} - V_{GS11}, \tag{3.89}$$

the minimal value of the common-mode input voltage will be given by the condition that transistors M1 and M2 be still biased in saturation region. In consequence, the minimal value of the common-mode input voltage will have the following expression:

$$V_{IC}^{\text{MIN}} = V_{D(M1)} + V_{SD1sat} - V_{SG1}, \tag{3.90}$$

equivalent to

$$V_{IC}^{\text{MIN}} = V_{D(M1)} - V_T. \tag{3.91}$$

Evaluating the expression of potential $V_{D(M1)}$ using Eq. (3.89) and replacing it in Eq. (3.91) result in

$$V_{IC}^{\text{MIN}} = V_{DD} - 5V_T - 3\sqrt{\frac{6I_{REF}}{K'_p\left(\frac{W}{L}\right)}} - \sqrt{\frac{I_{REF}}{K'_n\left(\frac{W}{L}\right)}}. \tag{3.92}$$

In order to determine the maximal value of the common-mode input voltage for the operational amplifier having the architecture presented in Fig. 3.5, suppose that the common-mode input voltage $v_{IN1} = v_{IN2} \stackrel{not}{=} V_{IC}$ is increasing. Because the source-gate voltages of transistors M1 and M2 are constant, the drain potential of transistor M3 will also decrease. The maximal value of this potential is given by the condition that transistor M3 be still biased in saturation region. In consequence, the maximal value of the common-mode input voltage will have the following expression:

$$V_{IC}^{\text{MAX}} = V_{DD} - V_{SD3sat} - V_{SG1}, \tag{3.93}$$

equivalent to

$$V_{IC}^{\text{MAX}} = V_{DD} - (V_{SG3} - V_T) - V_{SG1}, \tag{3.94}$$

or

$$V_{IC}^{\text{MAX}} = V_{DD} - V_T - \left(\sqrt{2} + 1\right)\sqrt{\frac{I_{REF}}{K'_p\left(\frac{W}{L}\right)}}. \tag{3.95}$$

References

1. Razavi B (2001) Design of Analog CMOS Integrated Circuits. New York: McGraw-Hill Education
2. Gray PR, Hurst PJ, Lewis SH and Meyer RG (2009) Analysis and Design of Analog Integrated Circuits, 5th ed. New York: Wiley
3. Allen PE and Holberg DR (2011) CMOS Analog Circuit Design, 3rd ed. New York: Oxford University Press
4. Sedra AS and Smith KC (2015) Microelectronic Circuits, 7th ed. New York: Oxford University Press

Chapter 4
CMOS Operational Amplifier Architectures: Fundamental Applications in Analog Circuit Design

Abstract Operational amplifiers are most commonly used in feedback configurations, where their behavior is determined primarily by external components rather than internal structure. When the open-loop gain is sufficiently high, these circuits can be accurately described using simplified relationships based mainly on passive component ratios. This chapter examines a wide range of operational amplifier applications under quasi-ideal feedback assumptions, while the internal amplifier is represented using one of the CMOS architectures introduced in the previous chapter. This approach separates system-level functionality from transistor-level implementation details. The analysis includes fundamental configurations such as voltage followers, inverting and noninverting amplifiers, difference amplifiers, high-gain stages and summing circuits, as well as more advanced structures like instrumentation amplifiers and precision rectifiers. For each circuit, analytical expressions are derived and validated through LTspice simulations. The sub-circuit representation of the operational amplifier ensures a direct link between theoretical analysis and simulation results. It is shown that, for sufficiently large open-loop gain, circuit performance is dominated by external component ratios, justifying simplified design equations. All applications may use any of the five CMOS operational amplifier architectures from Chap. 3, implemented through SPICE netlists included in the ". asc" files. Simulation files are provided as supplementary material and are intended for analysis, modification and validation in LTspice.

Keywords Linear and non-linear applications · Sub-circuit · Difference and high-gain amplifiers · Inverting voltage adder · Instrumentation amplifier · Precision rectifier

4.1 Introduction

Operational amplifiers are most commonly used in feedback configurations that define their functionality through external components rather than their internal structure. When the open-loop gain is sufficiently large, these circuits exhibit behavior that can be accurately described using simple relationships, often involving only ratios of passive components.

C. R. Popa, *Fundamentals of CMOS Analog Integrated Circuits*,
https://doi.org/10.1007/978-3-032-30677-7_4

In this chapter, a wide range of operational amplifier applications are analysed using quasi-ideal feedback assumptions, while the internal implementation of the amplifier is abstracted as one of the CMOS architectures introduced in the previous chapter. This approach allows the separation of system-level functionality from transistor-level complexity.

The chapter covers fundamental configurations such as voltage followers, inverting and noninverting amplifiers, difference and high-gain amplifiers, inverting voltage adders, as well as more advanced circuits including instrumentation amplifiers and precision rectifiers. For each application, analytical expressions are derived and validated through simulation.

The use of a sub-circuit representation for the operational amplifier enables seamless integration between theoretical analysis and LTspice simulation. The results confirm that, for sufficiently high open-loop gain, the circuit behavior is dominated by external component ratios, thereby justifying the use of simplified design formulas in practical applications.

For all the following applications using operational amplifiers, it is possible to use one of the five architectures of CMOS architectures of operational amplifiers, described in detail in Chap. 3. Each particular architecture has associated a netlist that must be inserted into the schematic (Edit/SPICE Directive/OK) for describing the particular architecture of the operational amplifier. In order to simplify the representation of all schematics from this chapter, this netlist is not shown, but it must exist in the ".asc" file whenever running a simulation.

All LTspice simulation files associated with the circuits presented in this book are provided as supplementary material on the publisher's website. Each file is named identically to the corresponding figure in the chapter, with the extension ".asc". The reader is encouraged to run, analyse and modify these files in order to validate the theoretical results and explore circuit behavior. The use of these materials requires the installation of the LTspice simulation software. The file OPAMP_DISCRET.asc must be placed in the same directory as all the other LTspice files in order to ensure proper simulation functionality.

All the following simulations are made in this work using **ARCHITECTURE 2 OF CMOS OPERATIONAL AMPLIFIER (Fig. 3.2 from Chap.** 3).

Netlist for Architecture 1 of CMOS Operational Amplifier (Fig. 3.1 from Chap. 3)

```
.SUBCKT OPAMP_DISCRET + - OUT VDD VSS
M7 N009 N009 VSS VSS cmosn l=1µm w=10µm
M4 VDD N002 N003 VDD cmosp l=1µm w=10µm
M8 N006 N009 VSS VSS cmosn l=1µm w=10µm
M9 N005 N009 VSS VSS cmosn l=1µm w=4.963µm
M1 N002 - N006 N006 cmosn l=1µm w=10µm
M2 N003 + N006 N006 cmosn l=1µm w=10µm
M3 VDD N002 N002 VDD cmosp l=1µm w=10µm
M5 VDD N003 N005 VDD cmosp l=1µm w=10µm
M6 VDD N005 OUT OUT cmosn l=1µm w=10µm
```

```
M10 OUT N009 VSS VSS cmosn l=1µm w=10µm
R N009 VDD 50kohm
.model cmosn NMOS (Kp=200µ Vto=0.5V Lambda=0.01)
.model cmosp PMOS (Kp=100µ Vto=-0.5V Lambda=0.02)
.ENDS OPAMP_DISCRET
```

Netlist for Architecture 2 of CMOS Operational Amplifier (Fig. 3.2 from Chap. 3)

```
.SUBCKT OPAMP_DISCRET + - OUT VDD VSS
R N010 N002 40kohm
M3 N011 N011 VSS VSS cmosn l=1µm w=10µm
M4 N012 N011 VSS VSS cmosn l=1µm w=10µm
M2 N003 + N012 N003 cmosp l=1µm w=10µm
M1 N003 - N011 N003 cmosp l=1µm w=10µm
M13 VDD N002 N003 VDD cmosp l=1µm w=10µm
M5 N005 N012 VSS VSS cmosn l=1µm w=10µm
M8 N006 N013 VSS VSS cmosn l=1µm w=10µm
M7 N013 N013 VSS VSS cmosn l=1µm w=10µm
M14 VDD N002 N005 VDD cmosp l=1µm w=5.133µm
M15 VDD N002 N006 VDD cmosp l=1µm w=10µm
M12 VDD N002 N002 VDD cmosp l=1µm w=10µm
M6 VDD N005 N013 VDD cmosp l=1µm w=10µm
M9 VDD N006 OUT OUT cmosn l=1µm w=10µm
M10 OUT N010 VSS VSS cmosn l=1µm w=10µm
M11 N010 N010 VSS VSS cmosn l=1µm w=10µm
.model cmosn NMOS (Kp=200µ Vto=0.5V Lambda=0.01)
.model cmosp PMOS (Kp=100µ Vto=-0.5V Lambda=0.02)
.ENDS OPAMP_DISCRET
```

Netlist for Architecture 3 of CMOS Operational Amplifier (Fig. 3.3 from Chap. 3)

```
.SUBCKT OPAMP_DISCRET + - OUT VDD VSS
M11 N020 N020 VSS VSS cmosn l=1µm w=10µm
M8 VDD N003 N005 VDD cmosp l=1µm w=10µm
M12 N015 N020 VSS VSS cmosn l=1µm w=10µm
M1 N009 - N015 N015 cmosn l=1µm w=10µm
M2 N010 + N015 N015 cmosn l=1µm w=10µm
M7 VDD N003 N003 VDD cmosp l=1µm w=10µm
M22 N019 N020 VSS VSS cmosn l=1µm w=5µm
R N020 N001 200kohm
M3 N006 N011 N009 N009 cmosn l=1µm w=10µm
M4 N007 N011 N010 N010 cmosn l=1µm w=10µm
M6 N005 N006 N007 N005 cmosp l=1µm w=10µm
M5 N003 N006 N006 N003 cmosp l=1µm w=10µm
M18 N008 N008 N012 N012 cmosn l=1µm w=10µm
```

```
M17 N004 N007 N008 N004 cmosp l=1µm w=10µm
M19 VDD N008 OUT OUT cmosn l=1µm w=10µm
M21 OUT N019 VSS OUT cmosp l=1µm w=10µm
M20 N012 N019 N019 N012 cmosp l=1µm w=10µm
M15 N018 N018 VSS VSS cmosn l=1µm w=1µm
M14 N014 N014 N018 N018 cmosn l=1µm w=1µm
M13 N011 N011 N014 N014 cmosn l=1µm w=1µm
M9 VDD N001 N001 VDD cmosp l=1µm w=10µm
M10 VDD N001 N011 VDD cmosp l=1µm w=10µm
M16 VDD N004 N004 VDD cmosp l=1µm w=10.05µm
.model cmosn NMOS (Kp=200µ Vto=0.5V Lambda=0.01)
.model cmosp PMOS (Kp=100µ Vto=-0.5V Lambda=0.02)
.ENDS OPAMP_DISCRET
```

Netlist for Architecture 4 of CMOS Operational Amplifier (Fig. 3.4 from Chap. 3)

```
.SUBCKT OPAMP_DISCRET + - OUT VDD VSS
M3 N010 N011 VSS VSS cmosn l=1µm w=10µm
M1 N003 + N010 N010 cmosn l=1µm w=10µm
M2 N004 - N010 N010 cmosn l=1µm w=10µm
M9 VDD N002 N004 VDD cmosp l=1µm w=10µm
M8 VDD N002 N003 VDD cmosp l=1µm w=10µm
M14 N009 N009 N014 N014 cmosn l=1µm w=10µm
M15 N008 N009 N015 N015 cmosn l=1µm w=10µm
M16 N014 N014 VSS VSS cmosn l=1µm w=10µm
M17 N015 N014 VSS VSS cmosn l=1µm w=10µm
M13 N004 N006 N008 N004 cmosp l=1µm w=10.008µm
M12 N003 N006 N009 N003 cmosp l=1µm w=10µm
M6 VDD N002 N011 VDD cmosp l=1µm w=10µm
M7 VDD N002 N002 VDD cmosp l=1µm w=10µm
M4 N011 N011 VSS VSS cmosn l=1µm w=10µm
M5 N002 N011 N018 N018 cmosn l=1µm w=40µm
R VSS N018 5.5kohm
M21 OUT N008 VSS OUT cmosp l=1µm w=10µm
M11 VDD N002 OUT VDD cmosp l=1µm w=10µm
M18 N006 N006 N007 N007 cmosn l=1µm w=10µm
M20 N016 N016 VSS VSS cmosn l=1µm w=10µm
M19 N007 N007 N016 N016 cmosn l=1µm w=10µm
M10 VDD N002 N006 VDD cmosp l=1µm w=30µm
.model cmosn NMOS (Kp=200µ Vto=0.5V Lambda=0.01)
.model cmosp PMOS (Kp=100µ Vto=-0.5V Lambda=0.02)
.ENDS OPAMP_DISCRET
```

Netlist for Architecture 5 of CMOS Operational Amplifier (Fig. 3.5 from Chap. 3)

```
.SUBCKT OPAMP_DISCRET + - OUT VDD VSS
M5 N016 N016 VSS VSS cmosn l=1µm w=10µm
R N016 N002 200kohm
M4 VDD N002 N002 VDD cmosp l=1µm w=10µm
M1 N014 - N010 N010 cmosp l=1µm w=10µm
M2 N013 + N010 N010 cmosp l=1µm w=10µm
M14 VDD N003 N003 VDD cmosp l=1µm w=10µm
M13 N005 N006 N009 N005 cmosp l=1µm w=10µm
M12 N003 N006 N006 N003 cmosp l=1µm w=10µm
M15 VDD N003 N005 VDD cmosp l=1µm w=10µm
M6 N013 N016 VSS VSS cmosn l=1µm w=10µm
M7 N014 N016 VSS VSS cmosn l=1µm w=10µm
M10 N006 N015 N013 N013 cmosn l=1µm w=10µm
M11 N009 N015 N014 N014 cmosn l=1µm w=9.995µm
M16 N007 N015 N015 N007 cmosp l=1µm w=10µm
M17 N004 N007 N007 N004 cmosp l=1µm w=10µm
M18 VDD N004 N004 VDD cmosp l=1µm w=10µm
M19 VDD N009 OUT OUT cmosn l=1µm w=10µm
M9 OUT N016 VSS VSS cmosn l=1µm w=10µm
M8 N015 N016 VSS VSS cmosn l=1µm w=30µm
M3 VDD N002 N010 VDD cmosp l=1µm w=10µm
.model cmosn nmos (Kp=200µ Vto=0.5V Lambda=0.01)
.model cmosp pmos (Kp=100µ Vto=-0.5V Lambda=0.02)
.ENDS OPAMP_DISCRET
```

After sizing the transistors of the operational amplifier and employing the circuit in several negative-feedback configurations (inverting and noninverting amplifiers, high-gain inverting amplifier, difference amplifier, instrumentation amplifier), a small output offset can be observed. This effect becomes more pronounced at high closed-loop gains. A closer analysis shows that the input differential stage of the operational amplifier does not operate perfectly symmetrically, even under ideal differential excitation. The main cause is the second-order effect of channel-length modulation, which introduces a dependence of the drain current on the drain-source voltage. In an ideal symmetric differential amplifier, the two transistors are expected to carry equal currents when the amplifier input voltage is zero. However, in the presence of channel-length modulation effect, any mismatch in the drain-source voltages between the two branches leads to slightly different drain currents, even when $v_{IN+} = v_{IN-}$. This current imbalance propagates through the subsequent stages and appears as an output offset, especially amplified in high closed-loop gain configurations. To mitigate this effect, a proper sizing of transistors, correlated with small adjustments of one aspect ratio *W/L*, is introduced for each architecture of operational amplifier. The goal of this adjustment is to equalize the drain-source voltages of the two transistors from the differential amplifier, $V_{DS1} = V_{DS2}$. Under this condition, the influence of the $1 + \lambda V_{DS}$ term becomes identical in both branches, restoring current symmetry and significantly reducing the offset. From a physical perspective, these modifications can be interpreted as compensating for channel-length modulation effect by rebalancing the DC operating points of the two transistors. Thus, although the circuit may appear geometrically symmetric, true electrical symmetry is only achieved after this correction. Simulation results confirm this analysis: after fine-tuning the aspect ratios, the output offset is significantly reduced, particularly in configurations with large closed-loop gain. In conclusion, geometrical symmetry does not guarantee electrical symmetry in the presence of second-order effects such as channel-length modulation.

The proper choosing and adjustments of aspect ratios W/L for each architecture of the operational amplifier are the following:

- The aspect ratio of transistor M9 for Architecture 1 of CMOS Operational Amplifier (Fig. 3.1 from Chap. 3), $(W/L)_9 = 5$ is replaced with $(W/L)_9 = 4.963$.
- The aspect ratio of transistor M14 for Architecture 2 of CMOS Operational Amplifier (Fig. 3.2 from Chap. 3), $(W/L)_{14} = 5$ is replaced with $(W/L)_{14} = 5.133$.
- The aspect ratio of transistor M16 for Architecture 3 of CMOS Operational Amplifier (Fig. 3.3 from Chap. 3), $(W/L)_{16} = 10$ is replaced with $(W/L)_{16} = 10.05$.
- The aspect ratio of transistor M13 for Architecture 4 of CMOS Operational Amplifier (Fig. 3.4 from Chap. 3), $(W/L)_{13} = 10$ is replaced with $(W/L)_{13} = 10.008$.
- The aspect ratio of transistor M11 for Architecture 5 of CMOS Operational Amplifier (Fig. 3.5 from Chap. 3), $(W/L)_{11} = 10$ is replaced with $(W/L)_{11} = 9.995$.

The closed-loop amplifier configurations presented in this chapter are based on standard operational amplifier theory [1–4].

4.2 Voltage Follower

The simplest application using an operational amplifier is represented by the voltage follower circuit (Fig. 4.1). Because of the very large value of the open-loop voltage gain of the operational amplifier, the output voltage of the circuit is practically identical to its input voltage:

$$v_O = v_{IN}. \tag{4.1}$$

Because the circuit has an infinite input impedance and a relatively low output impedance, it can be used for separating two blocks.

For graphically representing the dependence of the output voltage v_O on the input voltage v_{IN} for the voltage follower circuit from Fig. 4.1, a *DC sweep analysis* is performed in LTspice, for a variation of the input voltage v_{IN} between −200 and

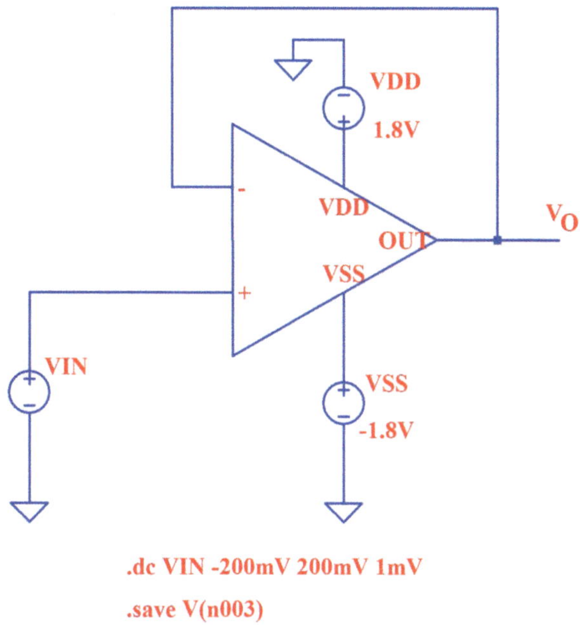

Fig. 4.1 Voltage follower circuit with a *DC sweep analysis* in LTspice

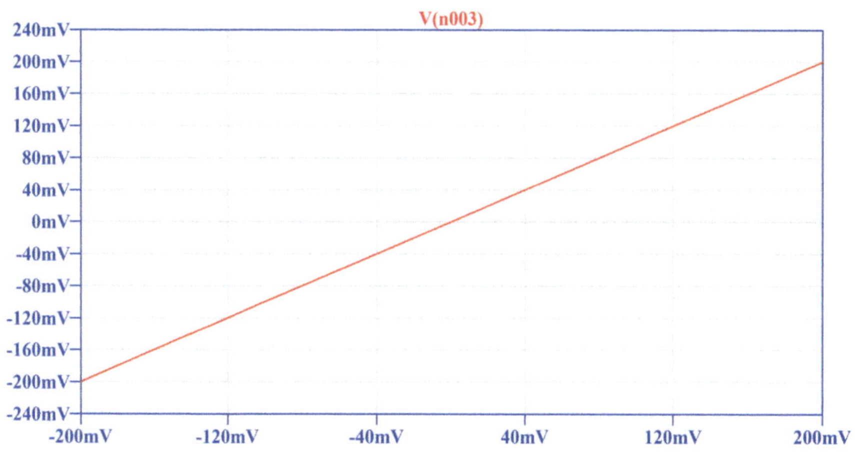

Fig. 4.2 Simulated characteristic $v_O(v_{IN})$ for the voltage follower

200 mV. In order to directly view the result of the simulation (the characteristic $v_O(v_{IN})$, shown in Fig. 4.2), the directive ".save V(n003)" is introduced in the file, with V(n003) representing the output voltage of the circuit v_O.

For graphically representing the dependence $v_O(t)$ for the voltage follower circuit from Fig. 4.3, a *transient analysis* is performed in LTspice, for a sinusoidal variation of the input voltage $v_{IN}(t)$. In order to directly view the result of the simulation (the

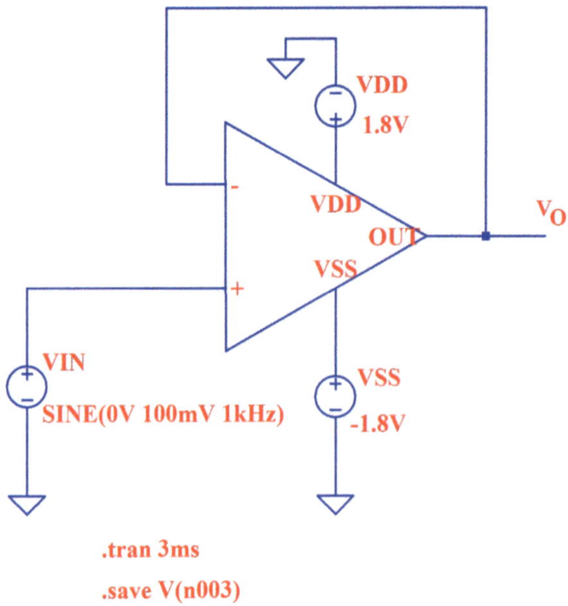

Fig. 4.3 Voltage follower circuit with a *transient analysis* in LTspice

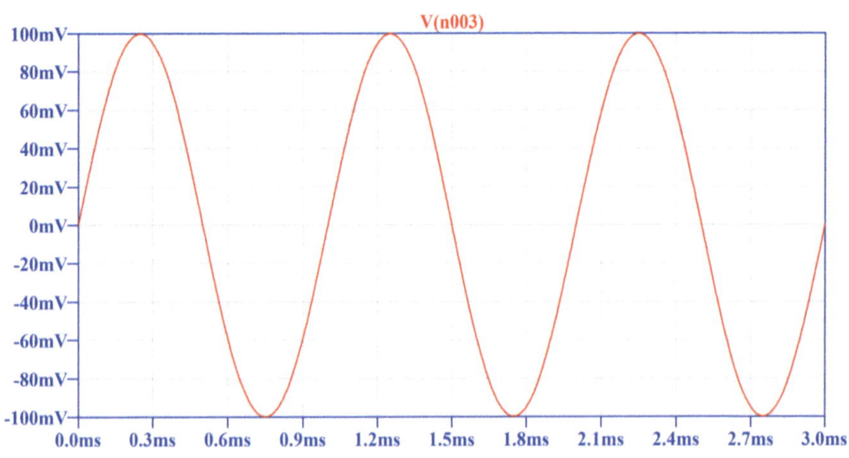

Fig. 4.4 Simulated characteristic $v_O(t)$ for the voltage follower

characteristic $v_O(t)$, shown in Fig. 4.4), the directive ".save V(n003)" is introduced in the file, with V(n003) representing the output voltage of the circuit v_O.

Both simulation results from Figs. 4.2 and 4.4 show an identity between the output voltage of the voltage follower circuit and its input voltage v_{IN}, confirming the theoretical estimated results.

4.3 Basic Inverting Amplifier

The basic inverting amplifier using an operational amplifier presented in Fig. 4.5 is able to offer a negative voltage gain, which is dependent on the ratio of two resistors, so it is practically independent of technologically caused errors and temperature variations.

Because the input current of the operational amplifier is neglectable, the currents which flow through resistors R_1 and R_2 are equal, so

$$\frac{v_{IN}}{R_1} = \frac{0 - v_O}{R_2}. \tag{4.2}$$

So, the closed-loop voltage gain of the inverting amplifier shown in Fig. 4.5 will have the following expression:

$$A = \frac{v_O}{v_{IN}} = -\frac{R_2}{R_1} = -\frac{100\,\mathrm{k\Omega}}{1\,\mathrm{k\Omega}} = -100. \tag{4.3}$$

For graphically representing the dependence of the output voltage v_O on the input voltage v_{IN} for the basic inverting amplifier from Fig. 4.5, a *DC sweep analysis* is performed in LTspice, for a variation of the input voltage v_{IN} between −2 and 2 mV. In order to directly view the result of the simulation (the characteristic $v_O(v_{IN})$, shown in Fig. 4.6), the directive ".save V(n002)" is introduced in the file, with V (n002) representing the output voltage of the circuit v_O.

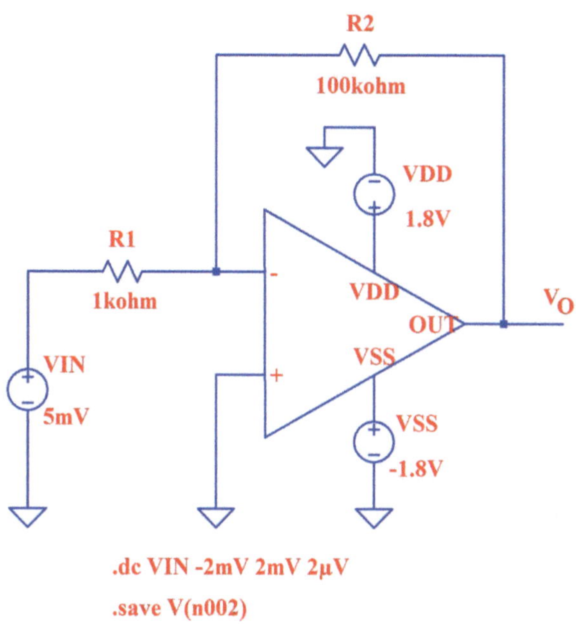

Fig. 4.5 Basic inverting amplifier with a *DC sweep analysis* in LTspice

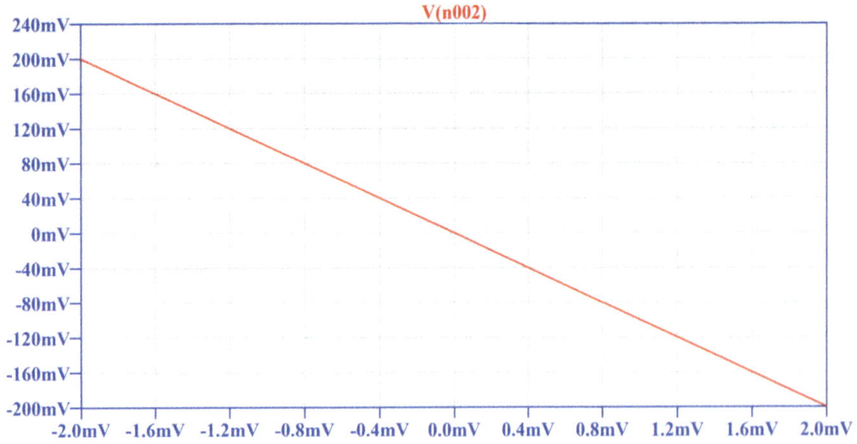

Fig. 4.6 Simulated characteristic $v_O(v_{IN})$ for the basic inverting amplifier

Fig. 4.7 Basic inverting amplifier with a *DC sweep analysis* in LTspice, having as parameter the resistance R_2

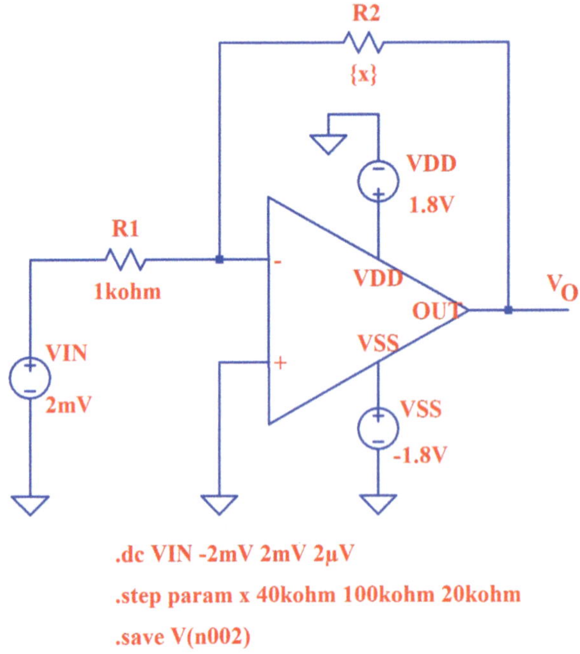

The slope of the simulated transfer characteristic from Fig. 4.6 is equal to -100, so the simulated voltage gain of the basic inverting amplifier presented in Fig. 4.5 is equal to its theoretical value, expressed by Eq. (4.3).

For illustrating the influence of the resistance R_2 on the operation of the basic inverting amplifier presented in Fig. 4.7, a *DC sweep analysis* is performed in LTspice, for a variation of the input voltage v_{IN} between -2 and $2\,\text{mV}$, and also a

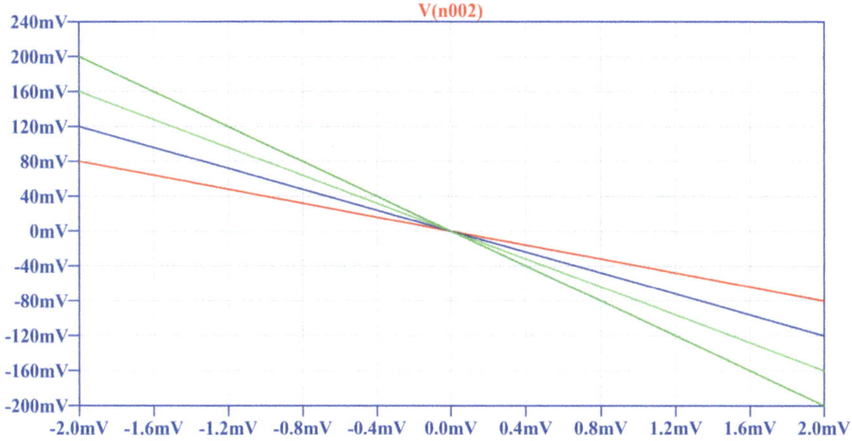

Fig. 4.8 Simulated characteristic $v_O(v_{IN})$ for the basic inverting amplifier, having as parameter the resistance R_2

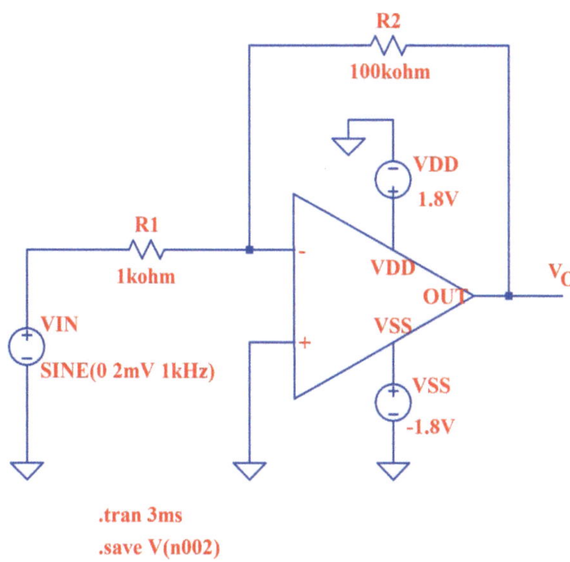

Fig. 4.9 Basic inverting amplifier circuit with a *transient analysis* in LTspice

parametrical simulation for $x = R_2$, between 40 and 100 kΩ, with a step of 20 kΩ. The circuit used for this simulation is presented in Fig. 4.7, and the simulation results are shown in Fig. 4.8. In order to directly view the results of the simulation, the directive ".save V(n002)" is introduced in the file, with V(n002) representing the output voltage of the circuit v_O.

For graphically representing the dependence $v_O(t)$ for the basic inverting amplifier from Fig. 4.9, a *transient analysis* is performed in LTspice, for a sinusoidal variation of the input voltage $v_{IN}(t)$, with an amplitude of 2 mV. In order to directly view the

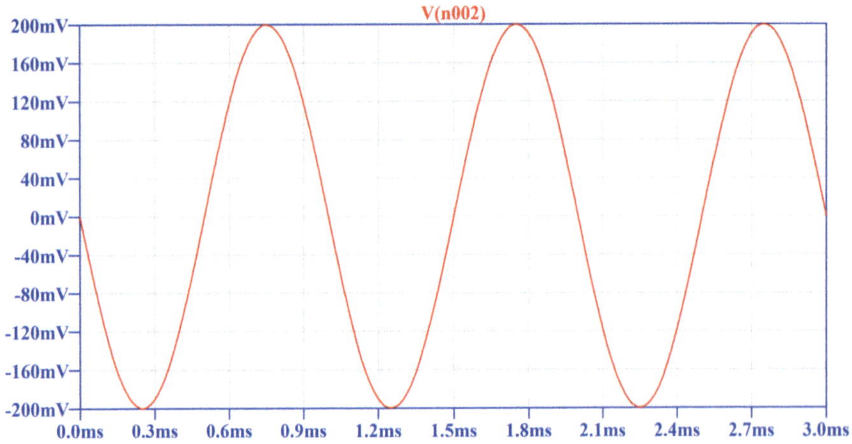

Fig. 4.10 Simulated characteristic $v_O(t)$ for the basic inverting amplifier

Fig. 4.11 Basic inverting amplifier with a *transient analysis* in LTspice, having as parameter the resistance R_2

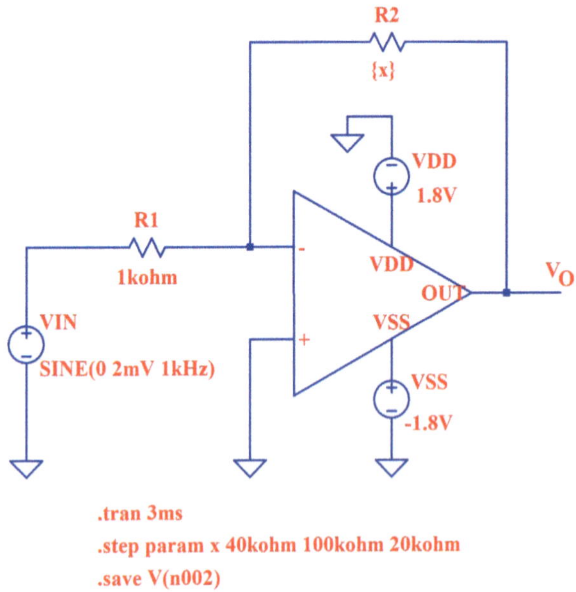

result of the simulation (the characteristic $v_O(t)$, shown in Fig. 4.10), the directive ". save V(n002)" is introduced in the file, with V(n002) representing the output voltage of the circuit v_O.

The amplitude of the simulated output voltage is 200 mV, so the simulated voltage gain is −100, identical to the theoretical voltage gain, expressed by Eq. (4.3).

For illustrating the influence of the resistance R_2 on the operation of the basic inverting amplifier presented in Fig. 4.11, a *transient analysis* is performed in LTspice, for a sinusoidal variation of the input voltage $v_{IN}(t)$, and also a parametrical

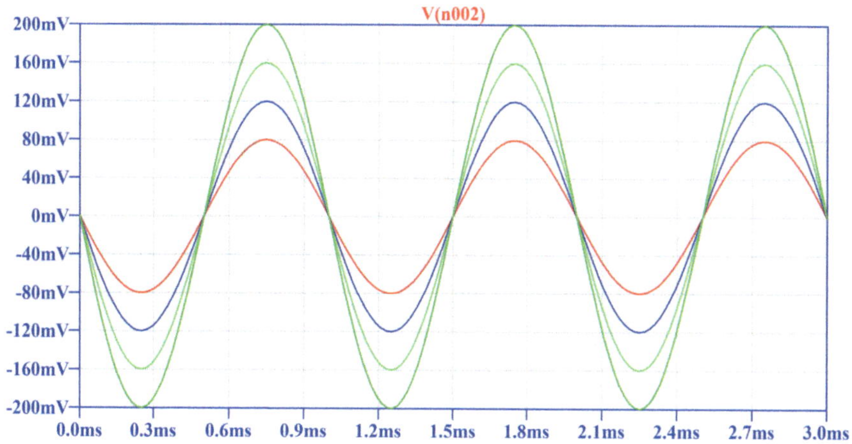

Fig. 4.12 Simulated characteristic $v_O(t)$ for the basic inverting amplifier, having as parameter the resistance R_2

simulation for $x = R_2$, between 40 and 100 kΩ, with a step of 20 kΩ. The circuit used for this simulation is presented in Fig. 4.11, and the simulation results are shown in Fig. 4.12. In order to directly view the results of the simulation, the directive ".save V(n002)" is introduced in the file, with V(n002) representing the output voltage of the circuit v_O.

4.4 High-Gain Inverting Amplifier

The requirement for the high-gain inverting amplifier is related to the difficulty of obtaining a high value of the voltage gain using the basic inverting amplifier from Fig. 4.1, because an important increasing of the absolute value of the voltage gain expressed by Eq. (4.3) is equivalent to the strong increasing of the value of resistor R_2. For example, a voltage gain equal to -1000 requires $R_1 = 1\,\text{k}\Omega$ and $R_2 = 1\,\text{M}\Omega$. Such a large value of resistor R_2 occupies an extremely large silicon area and, in consequence, must be avoided in usual designs. The minimal value of the resistor R_1 is limited by the requirements for minimal input resistance of the basic inverting amplifier. In conclusion, the maximal reasonable absolute value of the voltage gain for the basic inverting amplifier is of some hundreds.

The high-gain inverting amplifier presented in Fig. 4.13 solves this problem, having the possibility of obtaining an order of magnitude of the absolute value of its voltage gain of some thousands, by using resistors with reasonable values.

Noting with I_{R_k} the currents which flow through resistor R_k (having the sense from left to right), it is possible to write the following relations:

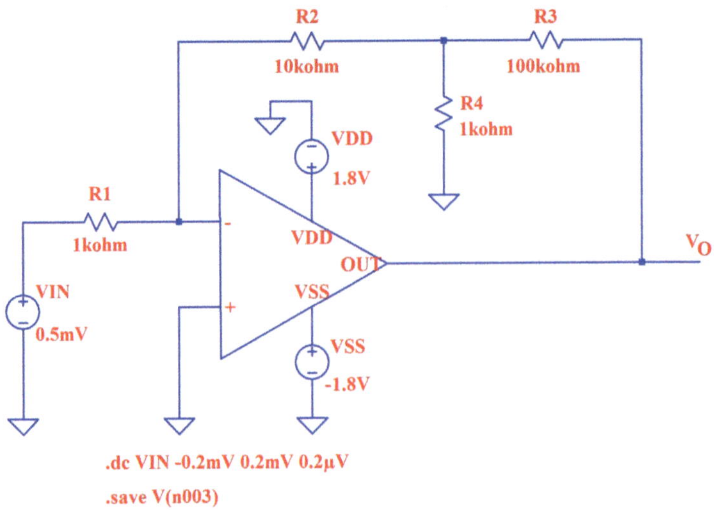

Fig. 4.13 High-gain inverting amplifier with a *DC sweep analysis* in LTspice

$$A = \frac{v_O}{v_{IN}} = \frac{v_O}{I_{R_3}} \frac{i_{R_3}}{i_{R_2}} \frac{i_{R_2}}{i_{R_1}} \frac{i_{R_1}}{v_{IN}}, \tag{4.4}$$

where the ratios that compose the previous relation can be expressed as follows:

$$\frac{v_O}{i_{R_3}} = -\left(R_3 + \frac{R_2 R_4}{R_2 + R_4}\right), \tag{4.5}$$

$$\frac{i_{R_2}}{i_{R_3}} = \frac{R_4}{R_2 + R_4}, \tag{4.6}$$

$$\frac{i_{R_2}}{i_{R_1}} = 1 \tag{4.7}$$

and

$$\frac{i_{R_1}}{v_{IN}} = \frac{1}{R_1}, \tag{4.8}$$

resulting in

$$A = -\frac{R_2 R_3 + R_2 R_4 + R_3 R_4}{R_1 R_4}. \tag{4.9}$$

Replacing the values of resistors from Fig. 4.13 results in

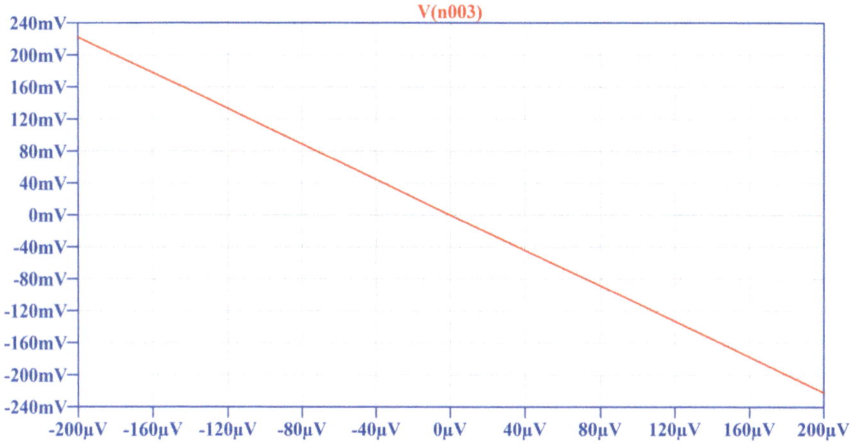

Fig. 4.14 Simulated characteristic $v_O(v_{IN})$ for the high-gain inverting amplifier

$$A = -\frac{10\,\mathrm{k\Omega} \times 1\,\mathrm{k\Omega} + 10\,\mathrm{k\Omega} \times 100\,\mathrm{k\Omega} + 1\,\mathrm{k\Omega} \times 100\,\mathrm{k\Omega}}{1\,\mathrm{k\Omega} \times 1\,\mathrm{k\Omega}} = -1110. \qquad (4.10)$$

For graphically representing the dependence of the output voltage v_O on the input voltage v_{IN} for the high-gain inverting amplifier from Fig. 4.13, a *DC sweep analysis* is performed in LTspice, for a variation of the input voltage v_{IN} between −0.2 and 0.2 mV. In order to directly view the result of the simulation (the characteristic $v_O(v_{IN})$, shown in Fig. 4.14), the directive ".save V(n003)" is introduced in the file, with V(n003) representing the output voltage of the circuit v_O.

The slope of the simulated transfer characteristic from Fig. 4.14 is equal to −1110, so the simulated voltage gain of the high-gain inverting amplifier presented in Fig. 4.13 is equal to its theoretical value, expressed by Eq. (4.10).

For illustrating the influence of the resistance R_1 on the operation of the high-gain inverting amplifier presented in Fig. 4.15, a *DC sweep analysis* is performed in LTspice, for a variation of the input voltage v_{IN} between −0.5 and 0.5 mV, and also a parametrical simulation for $x = R_1$, between 4 and 6 kΩ, with a step of 500 Ω. The circuit used for this simulation is presented in Fig. 4.15, and the simulation results are shown in Fig. 4.16. In order to directly view the results of the simulation, the directive ".save V(n003)" is introduced in the file, with V(n003) representing the output voltage of the circuit v_O.

For graphically representing the dependence $v_O(t)$ for the high-gain inverting amplifier from Fig. 4.17, a *transient analysis* is performed in LTspice, for a sinusoidal variation of the input voltage $v_{IN}(t)$, with an amplitude of 0.1 mV. In order to directly view the result of the simulation (the characteristic $v_O(t)$, shown in Fig. 4.18), the directive ".save V(n003)" is introduced in the file, with V(n003) representing the output voltage of the circuit v_O.

The amplitude of the simulated output voltage is 111 mV, so the simulated voltage gain is −1110, identical to the theoretical voltage gain, expressed by Eq. (4.10).

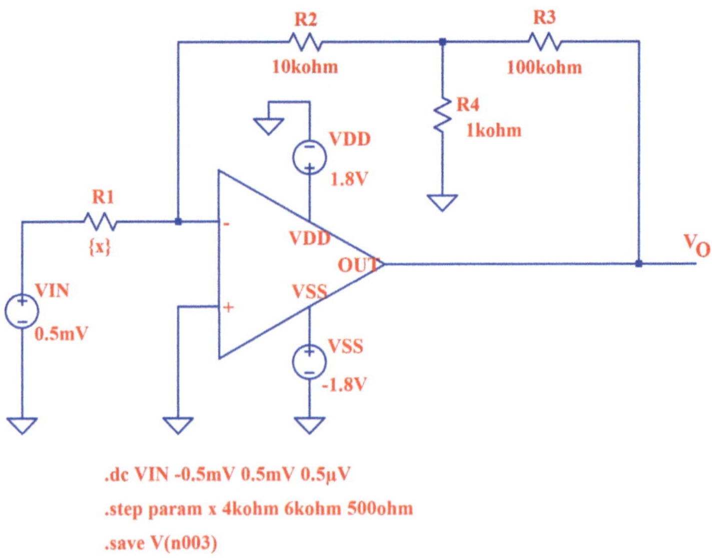

Fig. 4.15 High-gain inverting amplifier with a *DC sweep analysis* in LTspice, having as parameter the resistance R_1

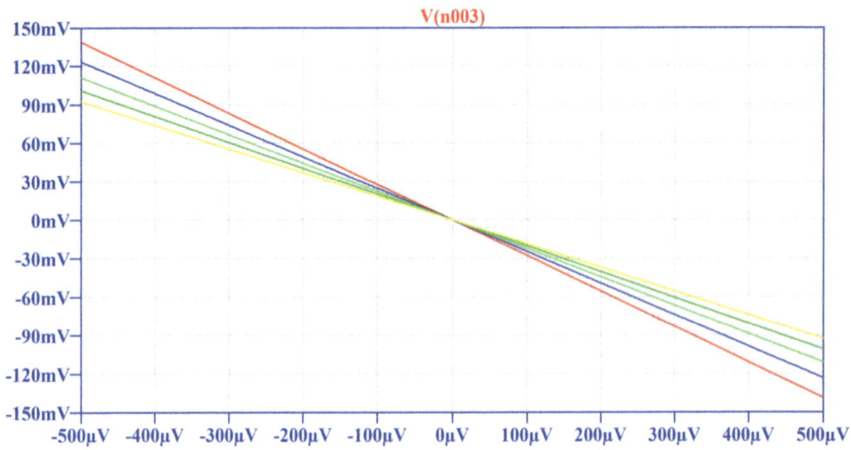

Fig. 4.16 Simulated characteristic $v_O(v_{IN})$ for the high-gain inverting amplifier, having as parameter the resistance R_1

For illustrating the influence of the resistance R_1 on the operation of the high-gain inverting amplifier presented in Fig. 4.19, a *transient analysis* is performed in LTspice, for a sinusoidal variation of the input voltage $v_{IN}(t)$ and also a parametrical simulation for $x = R_1$, between 4 and 6 kΩ, with a step of 500 Ω. The circuit used for

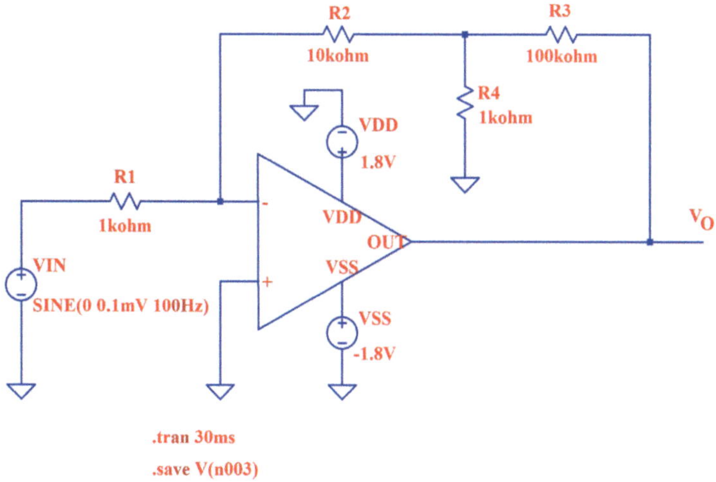

Fig. 4.17 High-gain inverting amplifier circuit with a *transient analysis* in LTspice

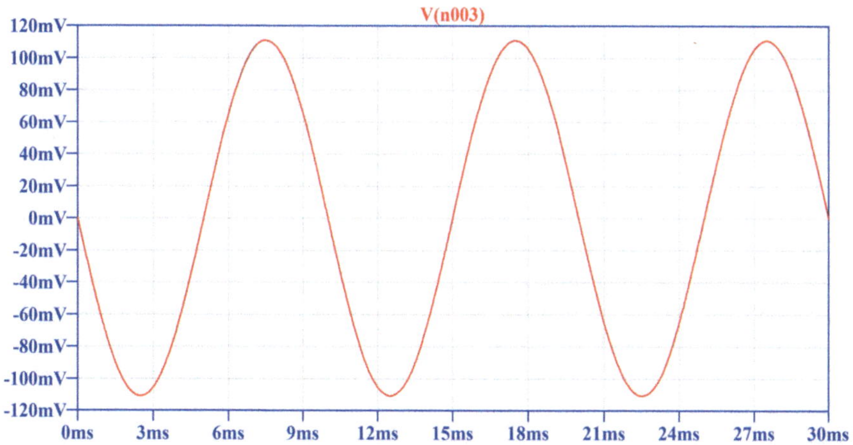

Fig. 4.18 Simulated characteristic $v_O(t)$ for the high-gain inverting amplifier

this simulation is presented in Fig. 4.19, and the simulation results are shown in Fig. 4.20. In order to directly view the results of the simulation, the directive ".save V(n003)" is introduced in the file, with V(n003) representing the output voltage of the circuit v_O.

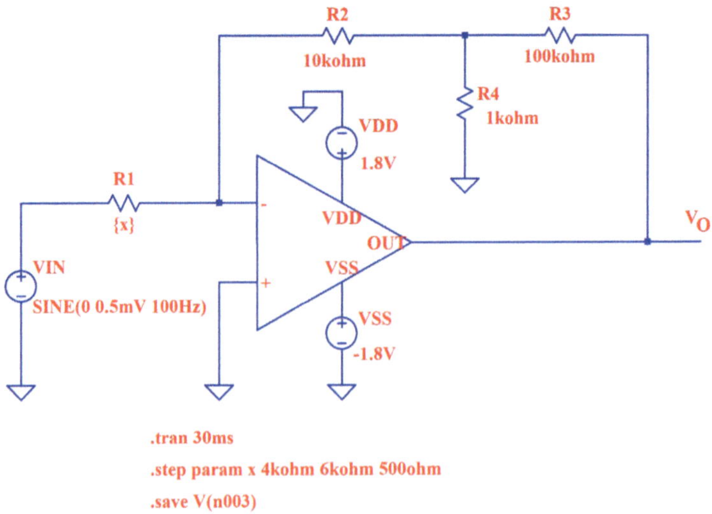

Fig. 4.19 High-gain inverting amplifier with a *transient analysis* in LTspice, having as parameter the resistance R_1

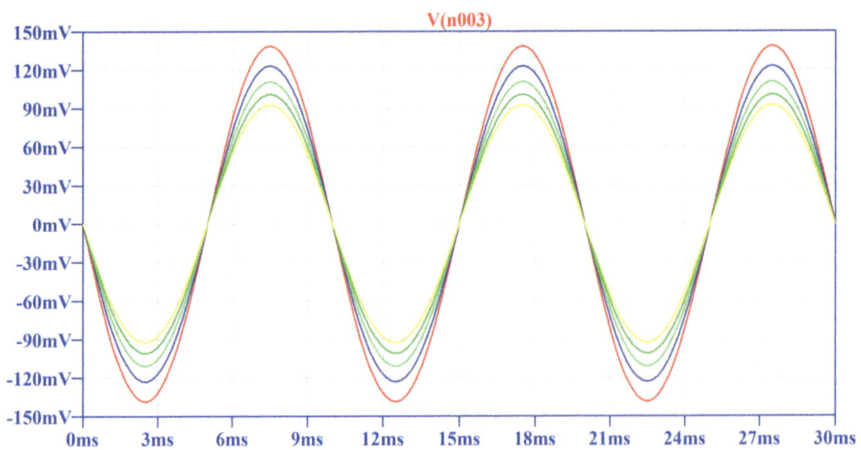

Fig. 4.20 Simulated characteristic $v_O(t)$ for the high-gain inverting amplifier, having as parameter the resistance R_1

4.5 Basic Noninverting Amplifier

In order to obtain positive controllable voltage gains, the basic noninverting amplifier circuit is presented in Fig. 4.21.

Because resistors R_1 and R_2 form a voltage divider, it is possible to write that

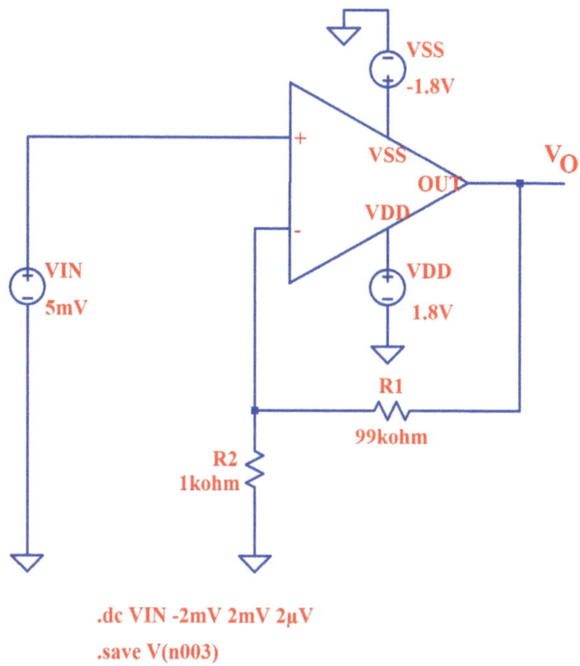

Fig. 4.21 Basic noninverting amplifier with a *DC sweep analysis* in LTspice

$$v_{IN} = v_O \frac{R_2}{R_1 + R_2}, \tag{4.11}$$

resulting in the following expression of the voltage gain for the basic noninverting amplifier presented in Fig. 4.21:

$$A = \frac{v_O}{v_{IN}} = 1 + \frac{R_2}{R_1} = 1 + \frac{99\,\text{k}\Omega}{1\,\text{k}\Omega} = 100. \tag{4.12}$$

For graphically representing the dependence of the output voltage v_O on the input voltage v_{IN} for the basic noninverting amplifier from Fig. 4.21, a *DC sweep analysis* is performed in LTspice, for a variation of the voltage v_{IN} between -2 and $2\,\text{mV}$. In order to directly view the result of the simulation (the characteristic $v_O(v_{IN})$, shown in Fig. 4.22), the directive ".save V(n003)" is introduced in the file, with V(n003) representing the output voltage of the circuit v_O.

The slope of the simulated transfer characteristic from Fig. 4.22 is equal to 100, so the simulated voltage gain of the basic noninverting amplifier presented in Fig. 4.21 is equal to its theoretical value, expressed by Eq. (4.12).

For illustrating the influence of the resistance R_2 on the operation of the basic noninverting amplifier presented in Fig. 4.23, a *DC sweep analysis* is performed in LTspice, for a variation of the voltage v_{IN} between -2 and $2\,\text{mV}$, and also a parametrical simulation for $x = R_2$, between 1 and $5\,\text{k}\Omega$, with a step of $1\,\text{k}\Omega$. The circuit used for this simulation is presented in Fig. 4.23, and the simulation results

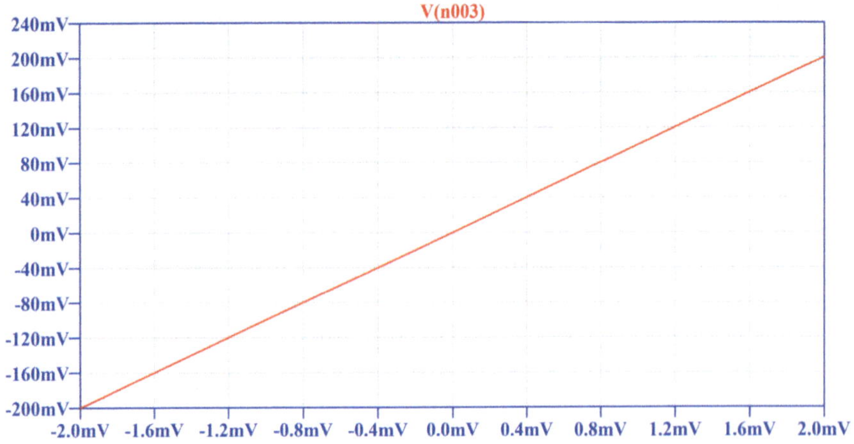

Fig. 4.22 Simulated characteristic $v_O(v_{IN})$ for the basic noninverting amplifier

Fig. 4.23 Basic noninverting amplifier with a *DC sweep analysis* in LTspice, having as parameter the resistance R_2

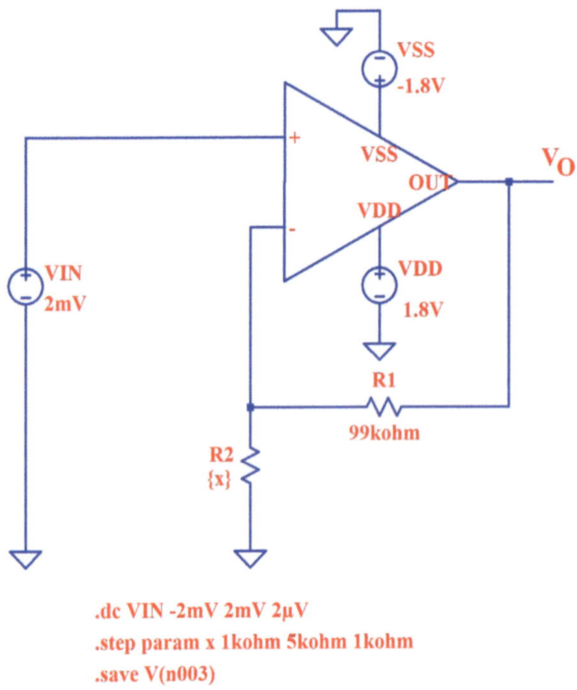

are shown in Fig. 4.24. In order to directly view the results of the simulation, the directive ".save V(n003)" is introduced in the file, with V(n003) representing the output voltage of the circuit v_O.

For graphically representing the dependence $v_O(t)$ for the basic noninverting amplifier from Fig. 4.25, a *transient analysis* is performed in LTspice, for a

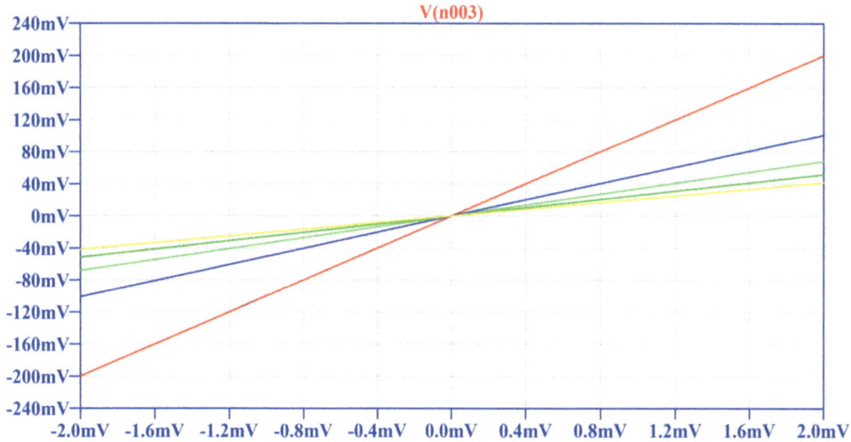

Fig. 4.24 Simulated characteristic $v_O(v_{IN})$ for the basic noninverting amplifier, having as parameter the resistance R_2

Fig. 4.25 Basic noninverting amplifier circuit with a *transient analysis* in LTspice

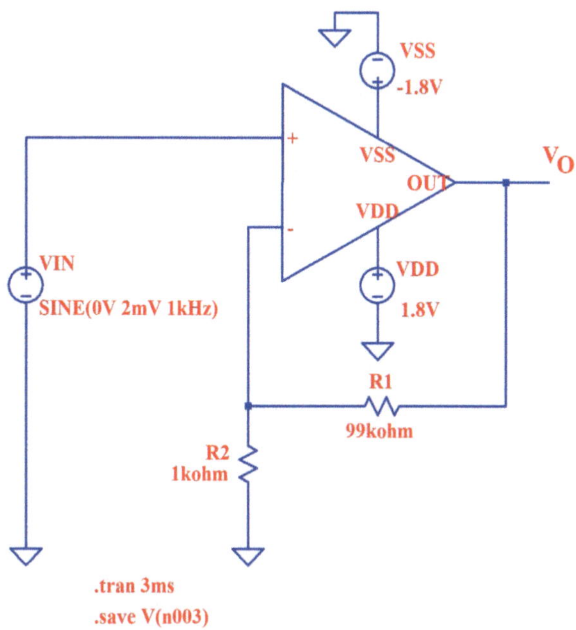

sinusoidal variation of the input voltage $v_{IN}(t)$, with an amplitude of 2 mV. In order to directly view the result of the simulation (the characteristic $v_O(t)$, shown in Fig. 4.26), the directive ".save V(n003)" is introduced in the file, with V(n003) representing the output voltage of the circuit v_O.

The amplitude of the simulated output voltage is 200 mV, so the simulated voltage gain is 100, identical to the theoretical voltage gain, expressed by Eq. (4.12).

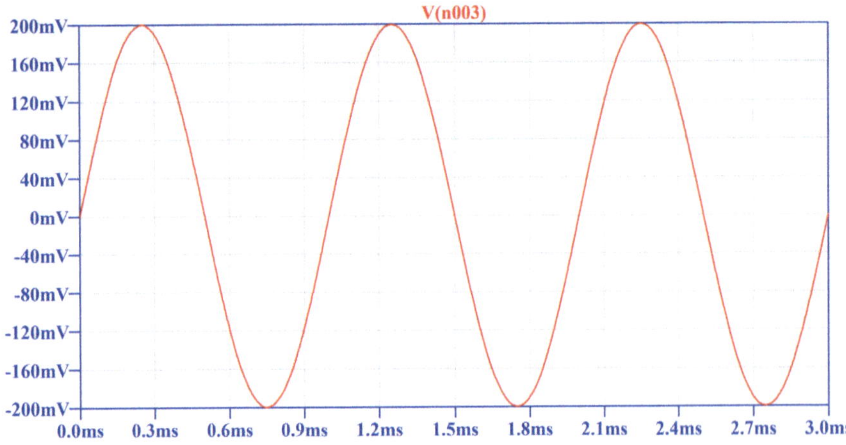

Fig. 4.26 Simulated characteristic $v_O(t)$ for the basic noninverting amplifier

Fig. 4.27 Basic noninverting amplifier with a *transient analysis* in LTspice, having as parameter the resistance R_2

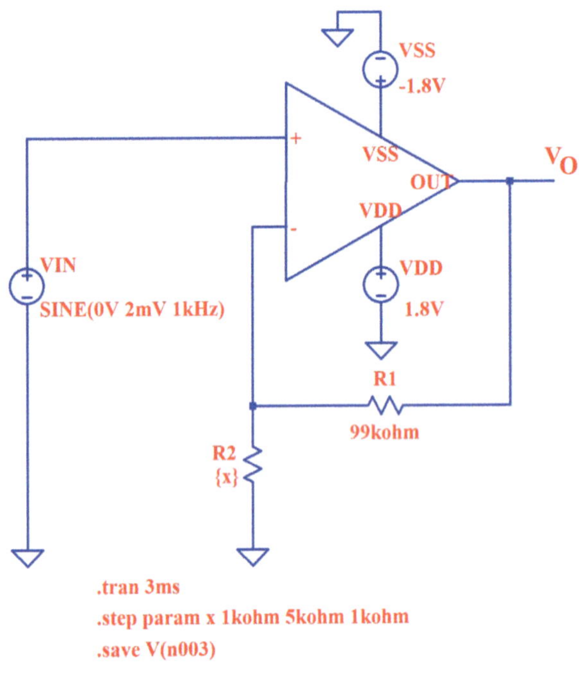

For illustrating the influence of the resistance R_2 on the operation of the nonamplifier presented in Fig. 4.27, a *transient analysis* is performed in LTspice, for a sinusoidal variation of the input voltage $v_{IN}(t)$, and also a parametrical simulation for $x = R_2$, between 1 and 5 kΩ, with a step of 1 kΩ. The circuit used for this simulation is presented in Fig. 4.27, and the simulation results are shown in Fig. 4.28. In order to directly view the results of the simulation, the directive ". save V(n003)" is introduced in the file, with V(n003) representing the output voltage of the circuit v_O.

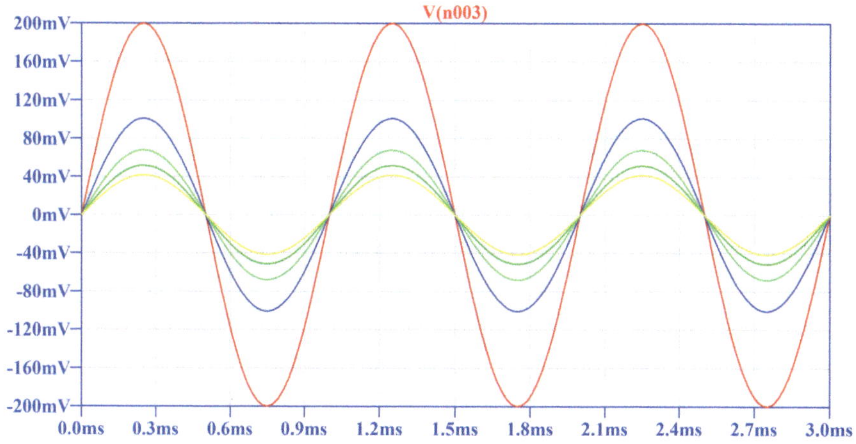

Fig. 4.28 Simulated characteristic $v_O(t)$ for the basic noninverting amplifier, having as parameter the resistance R_2

4.6 Inverting Voltage Adder

The inverting voltage adder circuit is presented in Fig. 4.29.

Using the superposition theorem, the output voltage v_O can be expressed as follows:

$$v_O = v_O|_{v_{IN1} \neq 0;\, v_{IN2} = 0} + v_O|_{v_{IN2} \neq 0;\, v_{IN1} = 0}. \tag{4.13}$$

Each term of the previous relation can be determined using the relation of the basic inverting amplifier (4.3):

$$v_O = \left(-\frac{R_3}{R_1}\right) v_{IN1} + \left(-\frac{R_3}{R_2}\right) v_{IN2}. \tag{4.14}$$

Replacing the values of resistors from Fig. 4.29 results in

$$v_O = -v_{IN1} - 10 v_{IN2}. \tag{4.15}$$

For graphically representing the dependence $v_O(t)$ for the inverting voltage adder from Fig. 4.29, a *transient analysis* is performed in LTspice, for a constant value of the first input voltage $v_{IN1} = 100\,\text{mV}$ and a sinusoidal variation of the second input voltage $v_{IN2}(t) = (10\,\text{mV})\sin(\omega t)$. Using Eq. (4.15), the theoretical expression of output voltage $v_O(t)$ will be

$$v_O(t) = -100\,\text{mV} - (100\,\text{mV})\sin(\omega t). \tag{4.16}$$

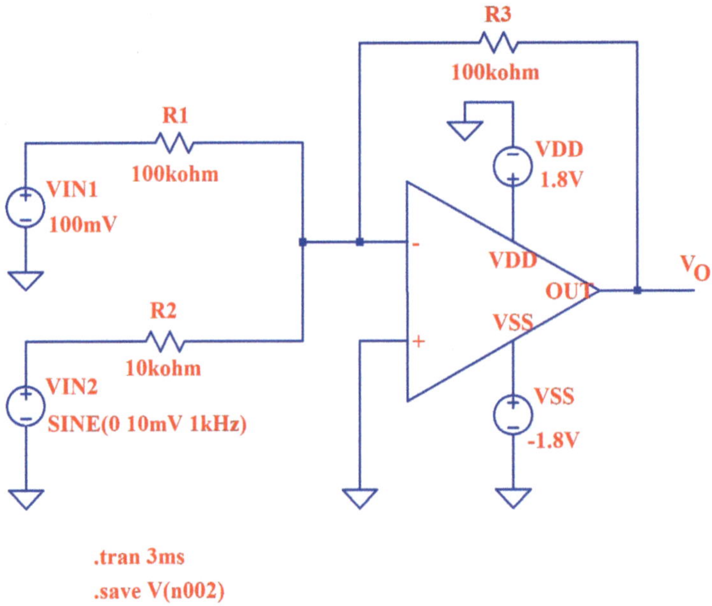

Fig. 4.29 Inverting voltage adder with a *transient analysis* in LTspice

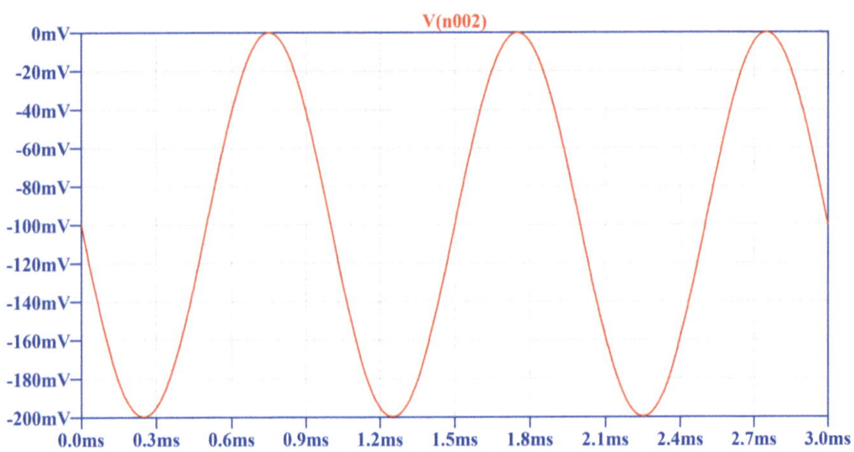

Fig. 4.30 Simulated characteristic $v_O(t)$ for the inverting voltage adder

In order to directly view the result of the simulation (the characteristic $v_O(t)$, shown in Fig. 4.30), the directive ".save V(n002)" is introduced in the file, with V (n002) representing the output voltage of the circuit v_O.

The simulated form of $v_O(t)$ dependence shown in Fig. 4.30 is identical to the theoretical dependence, expressed by Eq. (4.16).

For illustrating the influence of the first input voltage v_{IN1} on the operation of the inverting voltage adder presented in Fig. 4.31, a *transient analysis* is performed in

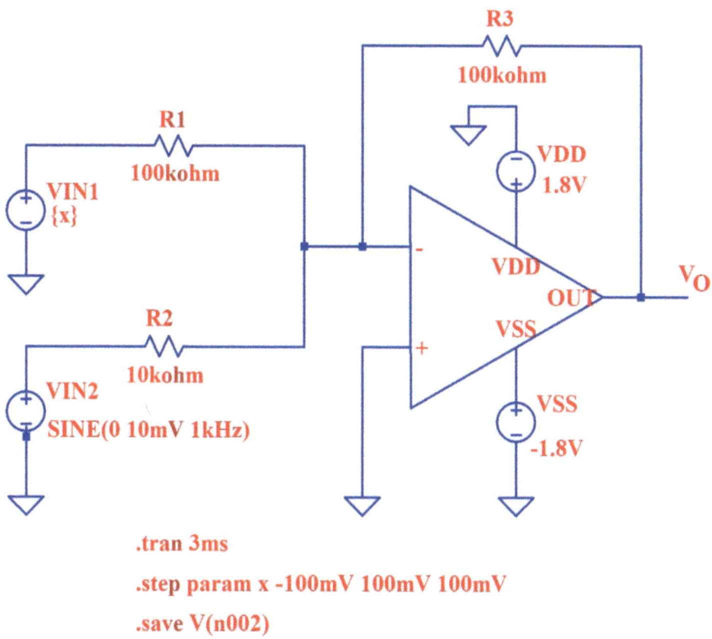

Fig. 4.31 Inverting voltage adder with a *transient analysis* in LTspice, having as parameter the first input voltage v_{IN1}

LTspice, for a sinusoidal variation of the second input voltage $v_{IN2}(t) = (10\,\text{mV})\sin(\omega t)$, and also a parametrical simulation for $x = v_{IN1}$, between -100 and 100 mV, with a step of 100 mV. The circuit used for this simulation is presented in Fig. 4.31, and the simulation results are shown in Fig. 4.32. In order to directly view the results of the simulation, the directive ".save V(n002)" is introduced in the file, with V(n002) representing the output voltage of the circuit v_O.

4.7 Basic Difference Amplifier

In order to compute the weight difference between two input voltages and also to amplify them, the basic difference amplifier circuit is presented in Fig. 4.33.

The operational amplifiers OA1 and OA2 are connected as voltage followers in order to separate the input voltage v_{ID} to the difference amplifier realized using the operational amplifier OA3 and resistors R_1, R_2, R_3 and R_4. Using the superposition theorem, the output voltage v_O can be expressed as follows:

$$v_O = v_O|_{v_{IN1} \neq 0;\, v_{IN2}=0} + v_O|_{v_{IN2} \neq 0;\, v_{IN1}=0}. \tag{4.17}$$

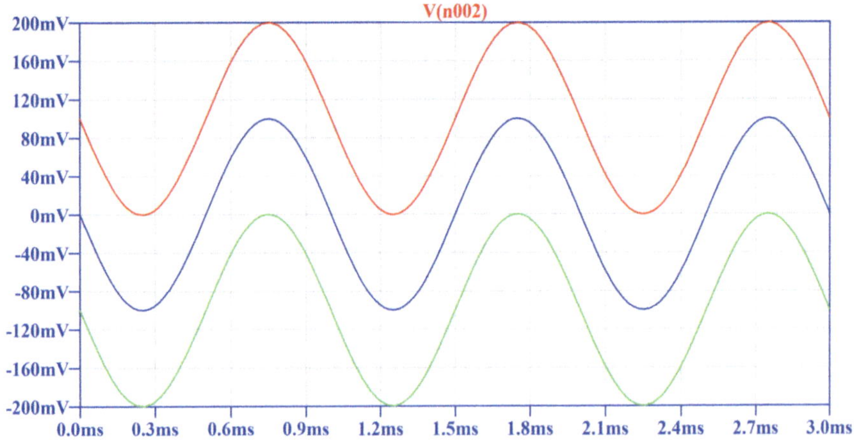

Fig. 4.32 Simulated characteristic $v_O(t)$ for the inverting voltage adder, having as parameter the first input voltage v_{IN1}

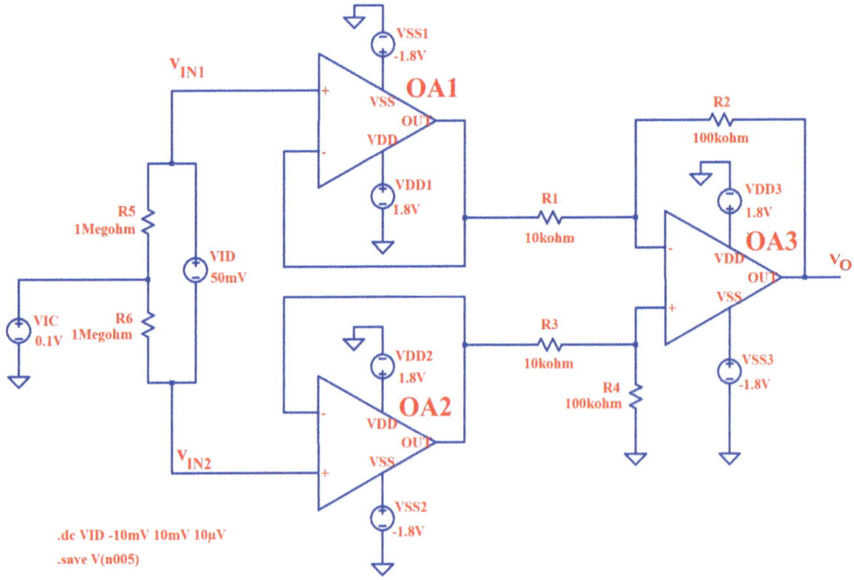

Fig. 4.33 Basic difference amplifier with a *DC sweep analysis* in LTspice

Because resistors R_3 and R_4 form a voltage divider, it is possible to write that

$$v_O = v_{IN1}\left(-\frac{R_2}{R_1}\right) + v_{IN2}\frac{R_4}{R_3 + R_4}\left(1 + \frac{R_2}{R_1}\right). \tag{4.18}$$

So, the general expression of the dependence of the output voltage v_O on the input voltages v_{IN1} and v_{IN2} is

$$v_O = A_1 v_{IN1} + A_2 v_{IN2}, \tag{4.19}$$

with $A_1 = -\frac{R_2}{R_1} < 0$ and $A_2 = \frac{R_4}{R_3+R_4}\left(1+\frac{R_2}{R_1}\right) > 0$ being the voltage gains for each input voltage. In general, their absolute values are not equal. In practical applications, the most convenient behavior of the basic difference amplifier from Fig. 4.33 is to be able to amplify the differential voltage $v_{ID} = v_{IN1} - v_{IN2}$. For this reason, usually the basic difference amplifier is designed with the condition that $|A_1| = A_2$, which is equivalent to

$$\frac{R_2}{R_1} = \frac{R_4}{R_3+R_4}\left(1+\frac{R_2}{R_1}\right), \tag{4.20}$$

resulting in the requirement for having the condition $R_1R_4 = R_2R_3$ fulfilled by a proper choosing of circuit resistors. In this particular hypothesis, the expression (4.18) of the output voltage v_O as a function on the input voltages v_{IN1} and v_{IN2} becomes

$$v_O = \frac{R_2}{R_1}(v_{IN2} - v_{IN1}) = -\frac{R_2}{R_1}v_{ID}, \tag{4.21}$$

with the basic difference amplifier from Fig. 4.33 being able to amplify (only in this particular scenario) the differential-mode input voltage v_{ID}.

In Fig. 4.33, the values of resistors are set in order to fulfil the previous condition, so the particular Eq. (4.21) becomes

$$v_O = -\frac{R_2}{R_1}v_{ID} = -\frac{100\,\text{k}\Omega}{10\,\text{k}\Omega}v_{ID} = -10v_{ID}. \tag{4.22}$$

For graphically representing the dependence of the output voltage v_O on the differential-mode input voltage v_{ID} for the basic difference amplifier from Fig. 4.33, a *DC sweep analysis* is performed in LTspice, for a variation of the voltage v_{ID} between -10 and 10 mV. In order to directly view the result of the simulation (the characteristic $v_O(v_{ID})$, shown in Fig. 4.34), the directive ".save V (n005)" is introduced in the file, with V(n005) representing the output voltage of the circuit v_O.

The slope of the simulated transfer characteristic from Fig. 4.34 is equal to -10, so the simulated voltage gain of the basic difference amplifier presented in Fig. 4.33 is equal to its theoretical value, expressed by Eq. (4.22).

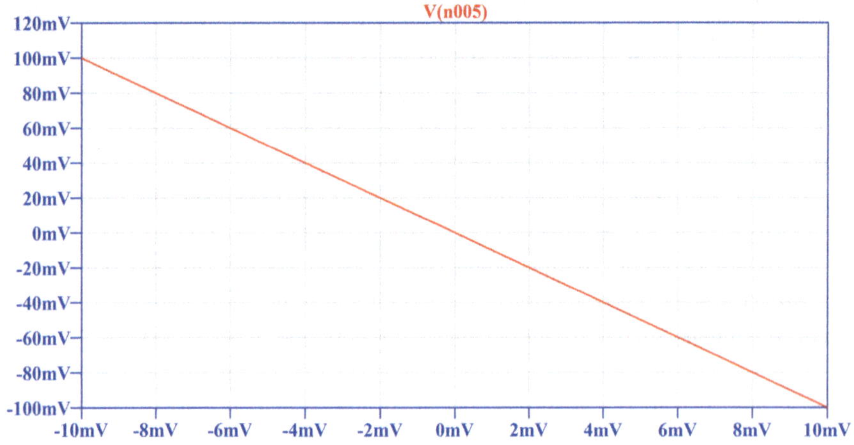

Fig. 4.34 Simulated characteristic $v_O(v_{ID})$ for the basic difference amplifier

For illustrating the influence of the resistances R_1 and R_3 on the operation of the basic difference amplifier presented in Fig. 4.35, a *DC sweep analysis* is performed in LTspice, for a variation of the voltage v_{ID} between -20 and 20 mV, and also a parametrical simulation for $x = R_1 = R_3$, between 10 and 40 kΩ, with a step of 10 kΩ. The circuit used for this simulation is presented in Fig. 4.35, and the simulation results are shown in Fig. 4.36. In order to directly view the results of the simulation, the directive ".save V(n005)" is introduced in the file, with V(n005) representing the output voltage of the circuit v_O.

For graphically representing the dependence $v_O(t)$ for the basic difference amplifier from Fig. 4.37, a *transient analysis* is performed in LTspice, for a sinusoidal variation of the input voltage $v_{ID}(t)$, with an amplitude of 20 mV. In order to directly view the result of the simulation (the characteristic $v_O(t)$, shown in Fig. 4.38), the directive ".save V(n005)" is introduced in the file, with V(n005) representing the output voltage of the circuit v_O.

The amplitude of the simulated output voltage is 200 mV, so the simulated voltage gain is -10, identical to the theoretical voltage gain, expressed by Eq. (4.22).

For illustrating the influence of the resistances $R_1 = R_3$ on the operation of the basic difference amplifier presented in Fig. 4.39, a *transient analysis* is performed in LTspice, for a sinusoidal variation of the input voltage $v_{ID}(t)$, and also a parametrical simulation for $x = R_1 = R_3$, between 10 and 40 kΩ, with a step of 10 kΩ. The circuit used for this simulation is presented in Fig. 4.39, and the simulation results are shown in Fig. 4.40. In order to directly view the results of the simulation, the directive ".save V(n005)" is introduced in the file, with V(n005) representing the output voltage of the circuit v_O.

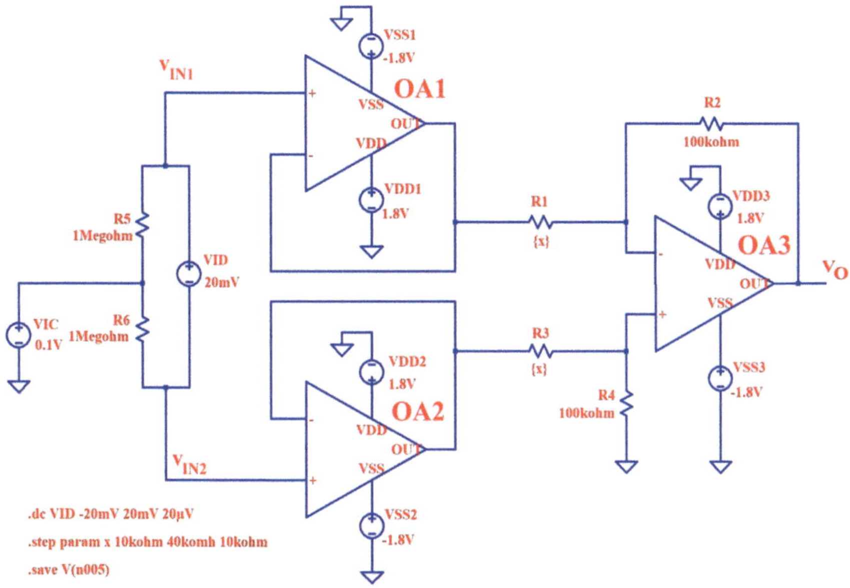

Fig. 4.35 Basic difference amplifier with a *DC sweep analysis* in LTspice, having as parameter the resistances $R_1 = R_3$

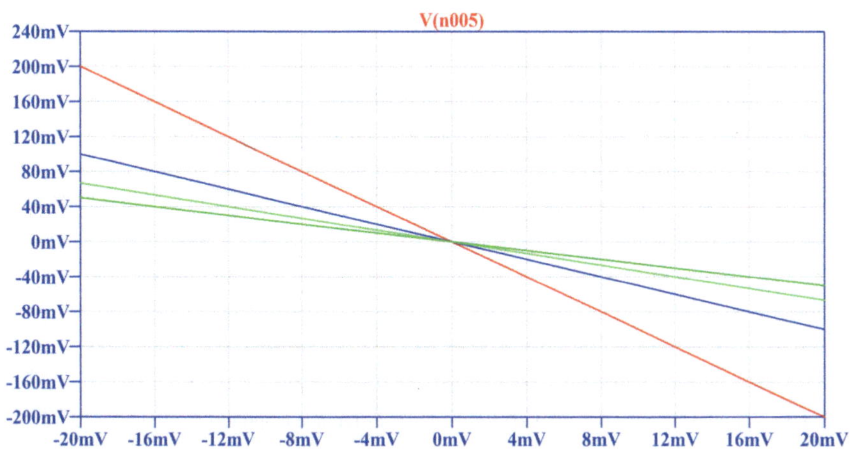

Fig. 4.36 Simulated characteristic $v_O(v_{ID})$ for the basic difference amplifier, having as parameter the resistances $R_1 = R_3$

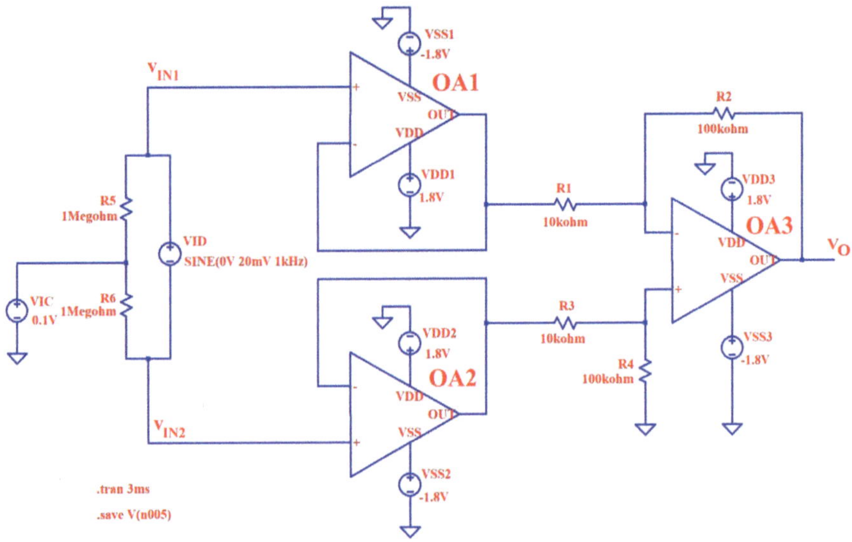

Fig. 4.37 Basic difference amplifier circuit with a *transient analysis* in LTspice

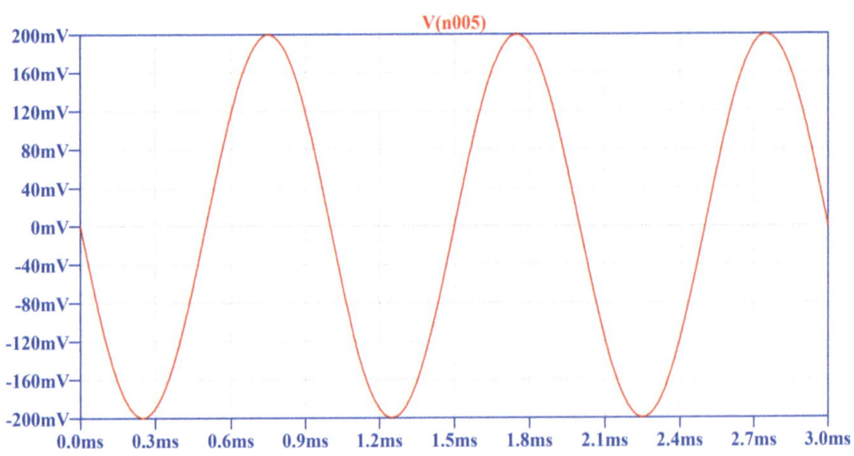

Fig. 4.38 Simulated characteristic $v_O(t)$ for the basic difference amplifier

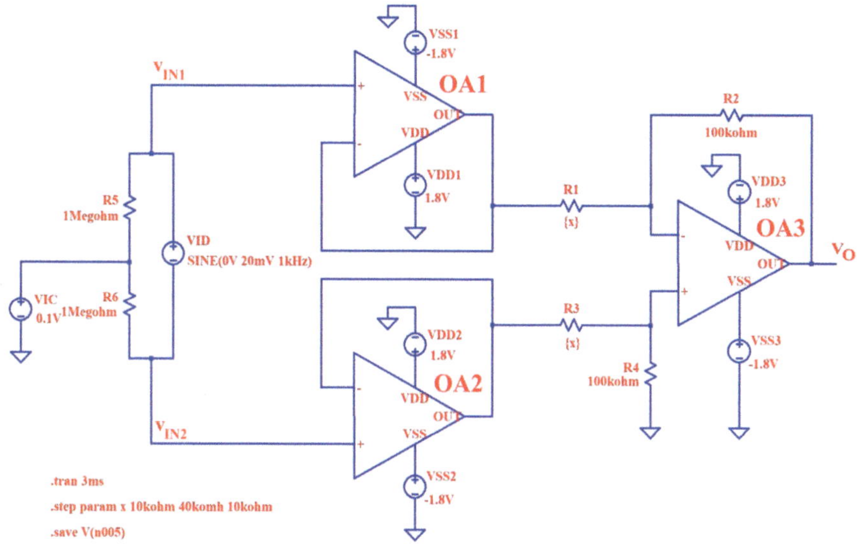

Fig. 4.39 Basic difference amplifier with a *transient analysis* in LTspice, having as parameter the resistances $R_1 = R_3$

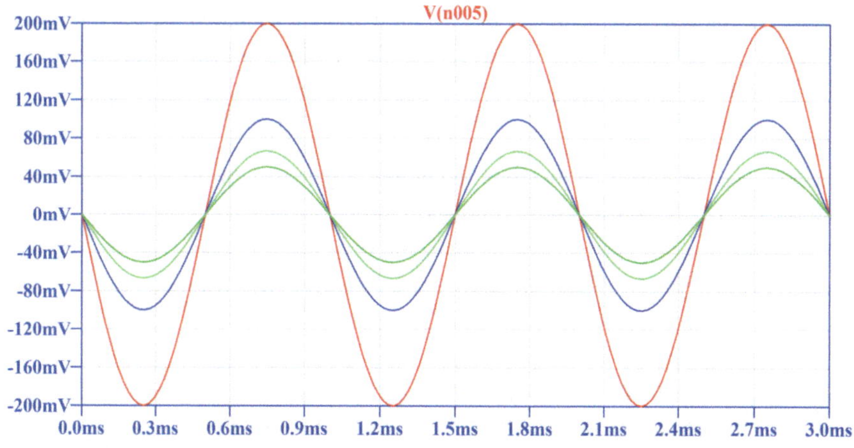

Fig. 4.40 Simulated characteristic $v_O(t)$ for the basic difference amplifier, having as parameter the resistances $R_1 = R_3$

4.8 Controlled Positive Feedback Difference Amplifier

For the controlled positive feedback difference amplifier presented in Fig. 4.41, the first and the second operational amplifiers OA1 and OA2 are connected as voltage follower circuits. They are used for offering an infinite input impedance to the circuit. In order to increase the voltage gain of the circuit, the difference amplifier

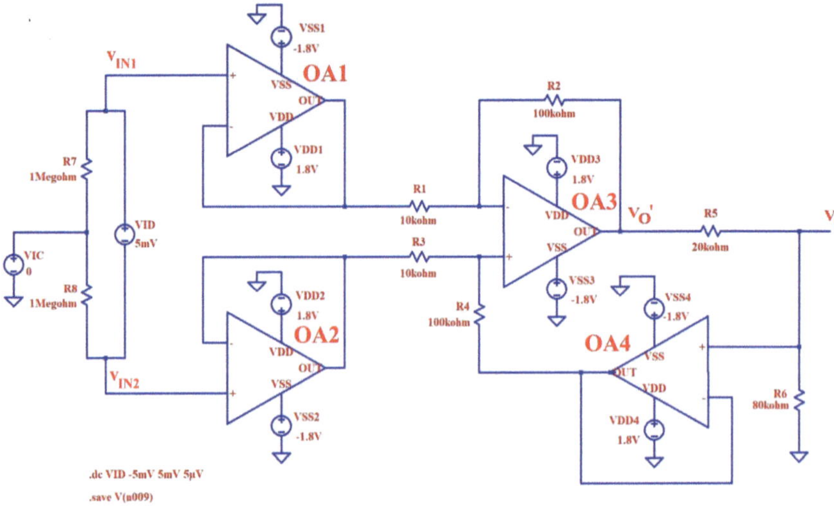

Fig. 4.41 Controlled positive feedback difference amplifier with *DC sweep analysis* in LTspice

presented in Fig. 4.41 uses a controlled positive feedback. The third operational amplifier OA3 works as a difference amplifier, but having the bottom terminal of resistor R_4 connected at the output of the fourth operational amplifier OA4, it is also used as a voltage follower. So, a part of the output voltage v_O is applied on the noninverting terminal of the opamp OA3, realizing a controllable positive feedback that increases the equivalent voltage gain of the entire difference amplifier from Fig. 4.41.

Using the superposition theorem, the output voltage $v_O^{'}$ of the third operational amplifier OA3 can be expressed as follows:

$$v_O{}' = v_O{}'|_{v_{IN1} \neq 0;\, v_{IN2}=0;\, v_O=0} + v_O{}'|_{v_{IN2} \neq 0;\, v_{IN1}=0;\, v_O=0} + v_O{}'|_{v_O \neq 0;\, v_{IN1}=0;\, v_{IN2}=0} \tag{4.23}$$

(because the output voltage v_O of the circuit becomes input voltage for the difference amplifier realized using the operational amplifier OA3, together with input voltages v_{IN1} and v_{IN2}). So, the output voltage $v_O^{'}$ of the operational amplifier OA3 will have the following expression:

$$v_O{}' = v_{IN1}\left(-\frac{R_2}{R_1}\right) + v_{IN2}\frac{R_4}{R_3+R_4}\left(1+\frac{R_2}{R_1}\right) + v_O\frac{R_3}{R_3+R_4}\left(1+\frac{R_2}{R_1}\right). \tag{4.24}$$

Resistors R_5 and R_6 form a voltage divider, so

$$v_O = v_O{}'\frac{R_6}{R_5+R_6}. \tag{4.25}$$

From Eqs. (4.24) and (4.25), the output voltage of the controlled positive feedback difference amplifier will be expressed by

$$v_O = \left(\frac{1 + \frac{R_2}{R_1}}{1 + \frac{R_3}{R_4}} v_{IN2} - \frac{R_2}{R_1} v_{IN1} \right) \left(1 + \frac{R_5}{R_6} - \frac{1 + \frac{R_2}{R_1}}{1 + \frac{R_4}{R_3}} \right)^{-1} . \tag{4.26}$$

So, the general expression of the dependence of the output voltage v_O on the input voltages v_{IN1} and v_{IN2} is

$$v_O = A_1 v_{IN1} + A_2 v_{IN2}, \tag{4.27}$$

with $A_1 < 0$ and $A_2 > 0$ being the voltage gains for each input voltage. In general, their absolute values are not equal. In practical applications, the most convenient behavior of the controlled positive feedback difference amplifier from Fig. 4.41 is to be able to amplify the differential voltage $v_{ID} = v_{IN1} - v_{IN2}$. For this reason, usually the controlled positive feedback difference amplifier is designed with the condition that $|A_1| = A_2$, which is equivalent to

$$\frac{1 + \frac{R_2}{R_1}}{1 + \frac{R_3}{R_4}} = \frac{R_2}{R_1}, \tag{4.28}$$

resulting in the requirement for having the condition $R_1 R_4 = R_2 R_3$ fulfilled by a proper choosing of circuit resistors. In this particular hypothesis, the expression (4.26) of the output voltage v_O as a function of the input voltages v_{IN1} and v_{IN2} becomes

$$v_O = \frac{R_2}{R_1} \frac{R_6}{R_5} (v_{IN2} - v_{IN1}) = -\frac{R_2}{R_1} \frac{R_6}{R_5} v_{ID}, \tag{4.29}$$

with the controlled positive feedback difference amplifier from Fig. 4.41 being able to amplify (only in this particular scenario) the differential-mode input voltage v_{ID}.

In Fig. 4.41, the values of resistors are set in order to fulfil the previous condition, so the particular Eq. (4.29) becomes

$$v_O = -\frac{R_2}{R_1} \frac{R_6}{R_5} v_{ID} = -\frac{100\,\text{k}\Omega}{10\,\text{k}\Omega} \frac{80\,\text{k}\Omega}{20\,\text{k}\Omega} v_{ID} = -40 v_{ID}. \tag{4.30}$$

For graphically representing the dependence of the output voltage v_O on the differential-mode input voltage v_{ID} for the controlled positive feedback difference amplifier from Fig. 4.41, a *DC sweep analysis* is performed in LTspice, for a variation of the voltage v_{ID} between −5 and 5 mV. In order to directly view the result of the simulation (the characteristic $v_O(v_{ID})$, shown in Fig. 4.42), the directive ".save V(n009)" is introduced in the file, with V(n009) representing the output voltage of the circuit v_O.

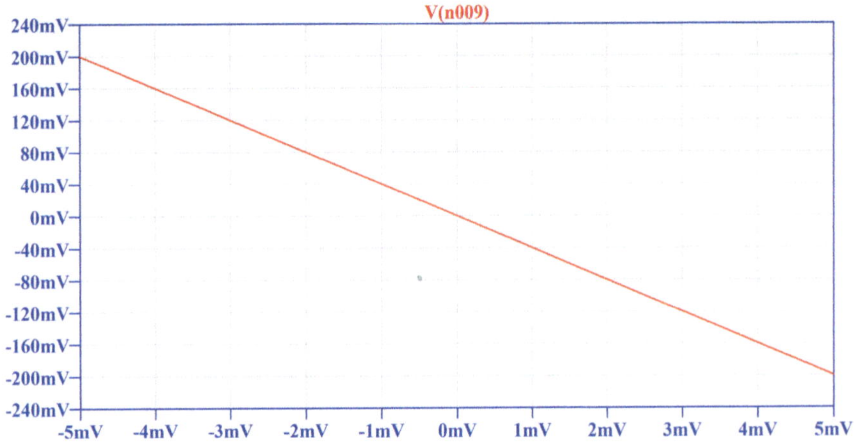

Fig. 4.42 Simulated characteristic $v_O(v_{ID})$ for the controlled positive feedback difference amplifier

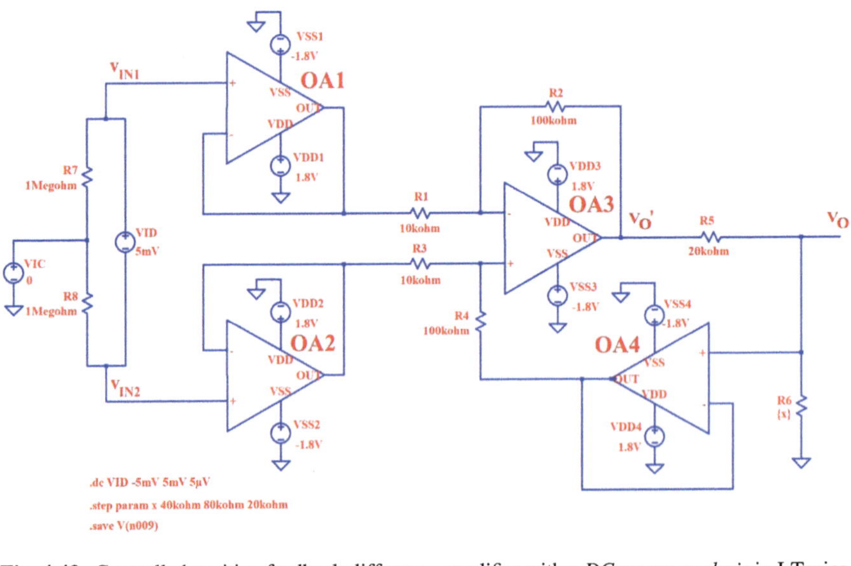

Fig. 4.43 Controlled positive feedback difference amplifier with a *DC sweep analysis* in LTspice, having as parameter the resistance R_6

The slope of the simulated transfer characteristic from Fig. 4.42 is equal to -40, so the simulated voltage gain of the controlled positive feedback difference amplifier presented in Fig. 4.41 is equal to its theoretical value, expressed by Eq. (4.30).

For illustrating the influence of the resistance R_6 on the operation of the controlled positive feedback difference amplifier presented in Fig. 4.43, a *DC sweep analysis* is performed in LTspice, for a variation of the voltage v_{ID} between -5 and 5 mV, and also a parametrical simulation for $x = R_6$, between 40 and 80 kΩ, with a step of 20 kΩ. The circuit used for this simulation is presented in Fig. 4.43, and the

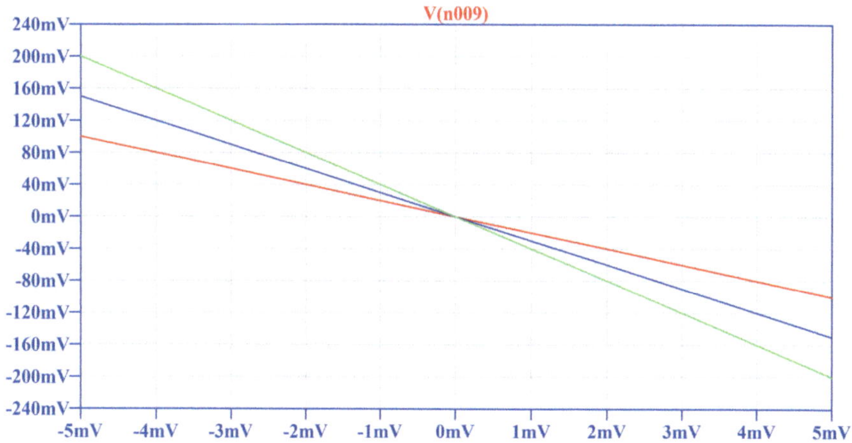

Fig. 4.44 Simulated characteristic $v_O(v_{ID})$ for the controlled positive feedback difference amplifier, having as parameter the resistance R_6

simulation results are shown in Fig. 4.44. In order to directly view the results of the simulation, the directive ".save V(n009)" is introduced in the file, with V(n009) representing the output voltage of the circuit v_O.

An interesting behavior of the controlled positive feedback difference amplifier from Fig. 4.43 is that the current which flows through resistor R_6 does not depend on the value of this resistor. Practically, the circuit operates like a voltage-controlled current source, with the current i_{R6} having the following expression:

$$i_{R_6} = -\frac{v_O}{R_6} = \frac{R_2}{R_1}\frac{1}{R_5}v_{ID}. \tag{4.31}$$

The sense of the current passing through resistor R_6 is set to be from the bottom to the top of this resistor. Replacing the values of resistors, the previous relation becomes

$$i_{R_6} = \frac{100\,\text{k}\Omega}{10\,\text{k}\Omega}\frac{1}{20\,\text{k}\Omega}v_{ID} = \frac{v_{ID}}{2\,\text{k}\Omega}. \tag{4.32}$$

For illustrating this behavior of the controlled positive feedback difference amplifier, the circuit presented in Fig. 4.45 is used. A *DC sweep analysis* is performed in LTspice, for a variation of the voltage v_{ID} between -5 and 5 mV, and also a parametrical simulation for $x = R_6$, between 40 and 100 kΩ, with a step of 20 kΩ. The circuit used for this simulation is presented in Fig. 4.45, and the simulation results are shown in Fig. 4.46. In order to directly view the results of the simulation, the directive ".save I(R6)" is introduced in the file, with I (R6) representing the current flowing through the resistor R_6.

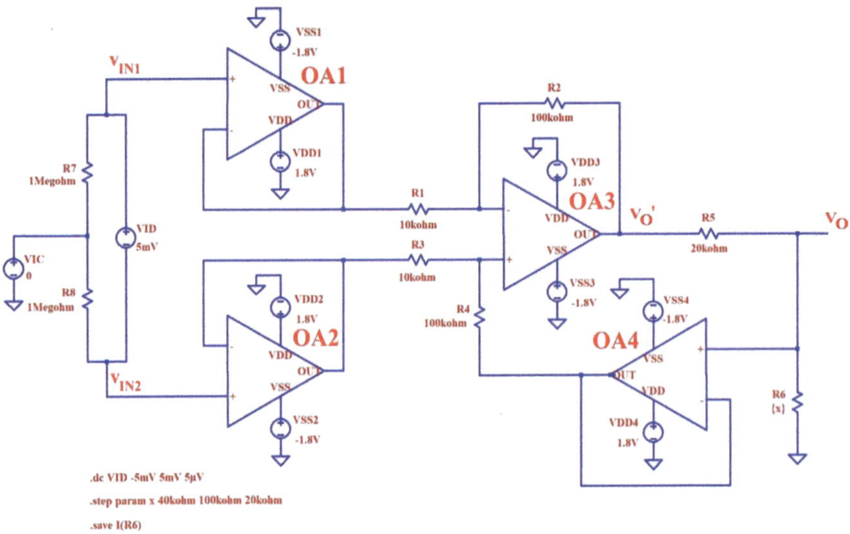

Fig. 4.45 Controlled positive feedback difference amplifier with a *DC sweep analysis* in LTspice, having as parameter the resistance R_6

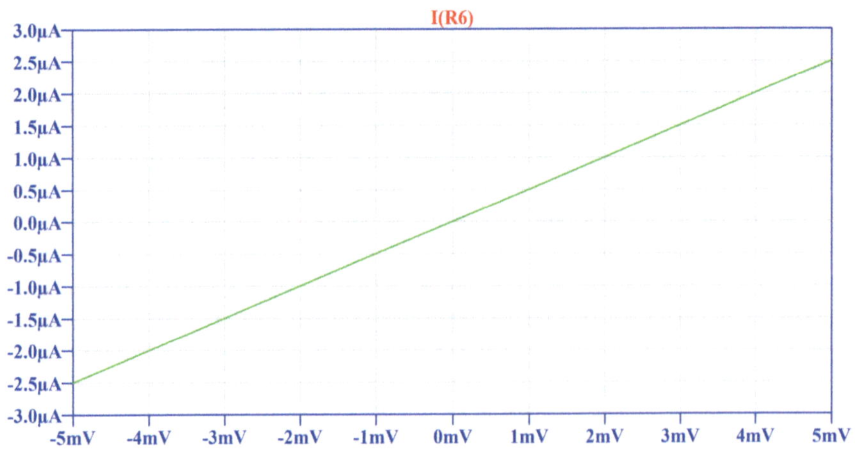

Fig. 4.46 Simulated characteristics $i_{R_6}(v_{ID})$ for the controlled positive feedback difference amplifier, having as parameter the resistance R_6

Although the simulation was carried out for multiple values of the resistor R_6, all resulting characteristics remain identical, confirming that the current i_{R6} is independent of the value of the resistor R_6. The slope of the characteristic from Fig. 4.46 is

$$tg(\alpha) = \frac{2.5\,\mu\text{A} - (-2.5\,\mu\text{A})}{5\,\text{mV} - (-5\,\text{mV})} = \frac{5\,\mu\text{A}}{10\,\text{mV}} = \frac{1}{2\,\text{k}\Omega}. \tag{4.33}$$

So, the simulated characteristic $i_{R_6}(v_{ID})$ will be

$$i_{R_6} = \frac{v_{ID}}{2\,\mathrm{k}\Omega}, \tag{4.34}$$

identical to the theoretical characteristic $i_{R_6}(v_{ID})$, expressed by Eq. (4.32).

4.9 Instrumentation Amplifier

In order to amplify with a large gain the differential-mode input voltage v_{ID} and to reject the common-mode input voltage v_{IC}, an instrumentation amplifier circuit can be used. A possible realization of such a circuit is presented in Fig. 4.47.

Using the superposition theorem for operational amplifiers OA1 and OA2, their output voltages v_{O1} and v_{O2} can be expressed as follows:

$$v_{O1} = v_{O1}\big|_{v_{IN1}\neq 0;\, v_{IN2}=0} + v_{O1}\big|_{v_{IN2}\neq 0;\, v_{IN1}=0} \tag{4.35}$$

and, respectively,

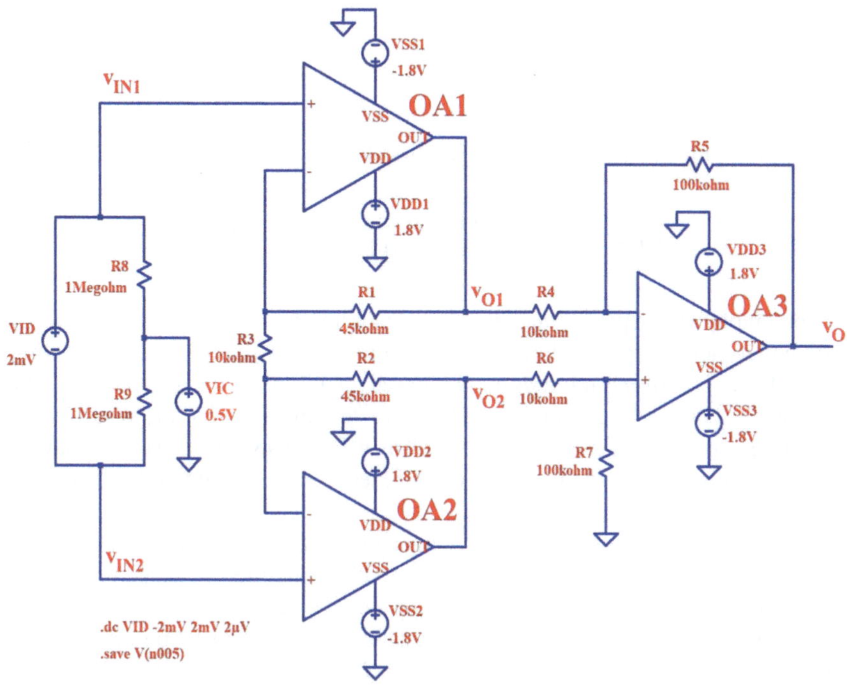

Fig. 4.47 Instrumentation amplifier with a *DC sweep analysis* in LTspice

$$v_{O2} = v_{O2}\big|_{v_{IN1}\neq 0;\, v_{IN2}=0} + v_{O2}\big|_{v_{IN2}\neq 0;\, v_{IN1}=0}, \tag{4.36}$$

equivalent to

$$v_{O1} = v_{IN1}\left(1+\frac{R_1}{R_3}\right) + v_{IN2}\left(-\frac{R_1}{R_3}\right) \tag{4.37}$$

and, respectively,

$$v_{O2} = v_{IN2}\left(1+\frac{R_2}{R_3}\right) + v_{IN1}\left(-\frac{R_2}{R_3}\right). \tag{4.38}$$

The third operational amplifier OA3 is used as a basic difference amplifier. The values of resistors R_4, R_5, R_6 and R_7 are set in order to fulfil the required condition for amplifying the differential input voltage, $v_{O2} - v_{O1}$ (see Eq. (4.21)):

$$v_O = \frac{R_5}{R_4}(v_{O2} - v_{O1}). \tag{4.39}$$

Replacing Eqs. (4.37) and (4.38) in Eq. (4.39) and using the identity of resistors R_1 and R_2 result in the following dependence of the output voltage v_O on the differential-mode input voltage v_{ID}:

$$v_O = \left(1+2\frac{R_1}{R_3}\right)\frac{R_5}{R_4}(v_{IN2} - v_{IN1}) = -\left(1+2\frac{R_1}{R_3}\right)\frac{R_5}{R_4}v_{ID}. \tag{4.40}$$

Replacing the values of the resistors, it results that

$$v_O = -\left(1+2\frac{45\,\text{k}\Omega}{10\,\text{k}\Omega}\right)\frac{100\,\text{k}\Omega}{10\,\text{k}\Omega}v_{ID} = -100v_{ID}. \tag{4.41}$$

For graphically representing the dependence of the output voltage v_O on the differential-mode input voltage v_{ID} for the instrumentation amplifier from Fig. 4.47, a *DC sweep analysis* is performed in LTspice, for a variation of the voltage v_{ID} between -2 and 2 mV. In order to directly view the result of the simulation (the characteristic $v_O(v_{ID})$, shown in Fig. 4.48), the directive ".save V (n005)" is introduced in the file, with V(n005) representing the output voltage of the circuit v_O.

The slope of the simulated transfer characteristic from Fig. 4.48 is equal to -100, so the simulated voltage gain of the instrumentation amplifier presented in Fig. 4.47 is equal to its theoretical value, expressed by Eq. (4.41).

For illustrating the influence of the resistance R_3 on the operation of the instrumentation amplifier presented in Fig. 4.49, a *DC sweep analysis* is performed in LTspice, for a variation of the voltage v_{ID} between -2 and 2 mV, and also a

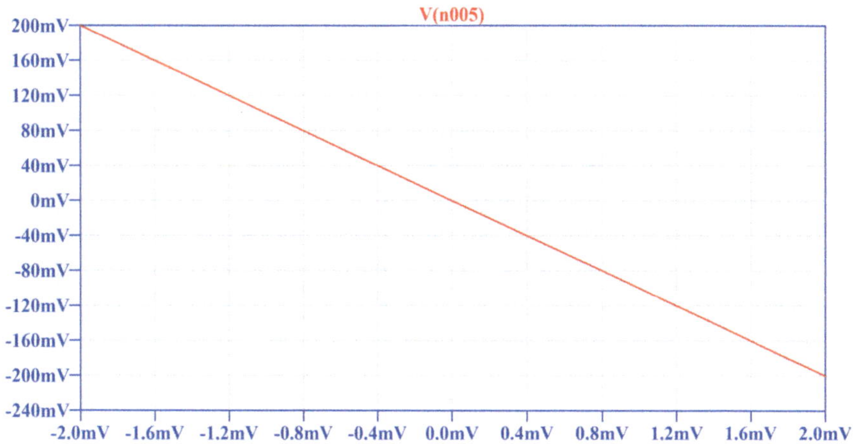

Fig. 4.48 Simulated characteristic $v_O(v_{ID})$ for the instrumentation amplifier

parametrical simulation for $x = R_3$, between 10 and 25 kΩ, with a step of 5 kΩ. The circuit used for this simulation is presented in Fig. 4.49, and the simulation results are shown in Fig. 4.50. In order to directly view the results of the simulation, the directive ".save V(n005)" is introduced in the file, with V(n005) representing the output voltage of the circuit v_O.

For graphically representing the dependence $v_O(t)$ for the instrumentation amplifier from Fig. 4.51, a *transient analysis* is performed in LTspice, for a sinusoidal variation of the input voltage $v_{ID}(t)$, with an amplitude of 2 mV. In order to directly view the result of the simulation (the characteristic $v_O(t)$, shown in Fig. 4.52), the directive ".save V(n005)" is introduced in the file, with V(n005) representing the output voltage of the circuit v_O.

The amplitude of the simulated output voltage is 200 mV, so the simulated voltage gain is -100, identical to the theoretical voltage gain, expressed by Eq. (4.41).

For illustrating the influence of the resistance R_3 on the operation of the instrumentation amplifier presented in Fig. 4.53, a *transient analysis* is performed in LTspice, for a sinusoidal variation of the input voltage $v_{ID}(t)$, and also a parametrical simulation for $x = R_3$, between 10 and 25 kΩ, with a step of 5 kΩ. The circuit used for this simulation is presented in Fig. 4.53, and the simulation results are shown in Fig. 4.54. In order to directly view the results of the simulation, the directive ".save V(n005)" is introduced in the file, with V(n005) representing the output voltage of the circuit v_O.

As it can be observed from previous theoretical analysis and also by simulation results, the common-mode input voltage $v_{IC} = 0.5$ V is totally rejected by the instrumentation amplifier.

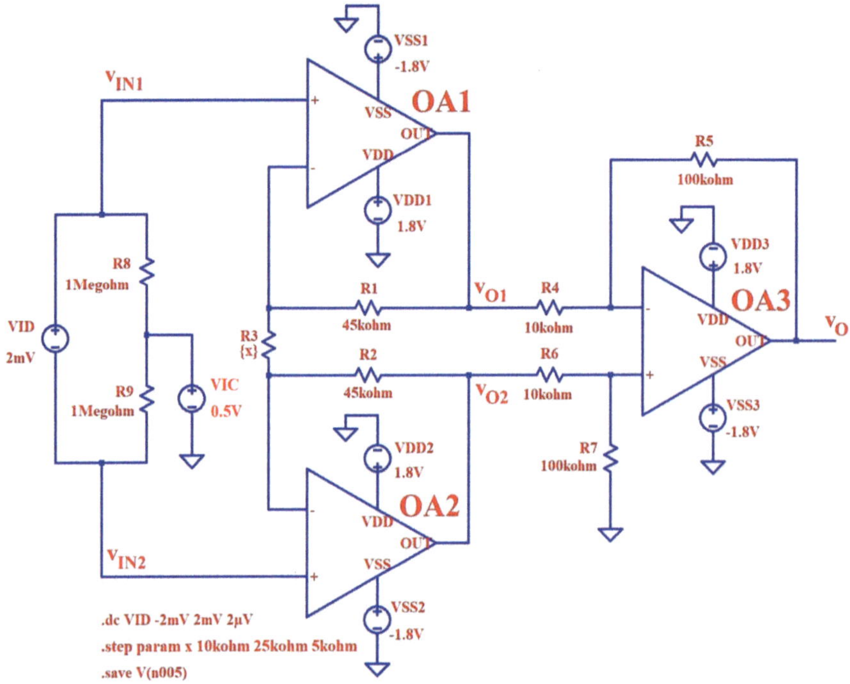

Fig. 4.49 Instrumentation amplifier with a *DC sweep analysis* in LTspice, having as parameter the resistance R_3

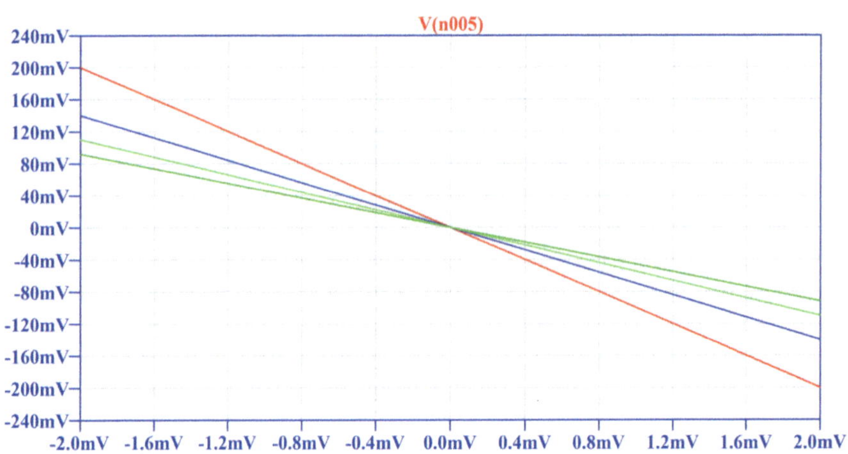

Fig. 4.50 Simulated characteristic $v_O(v_{ID})$ for the instrumentation amplifier, having as parameter the resistance R_3

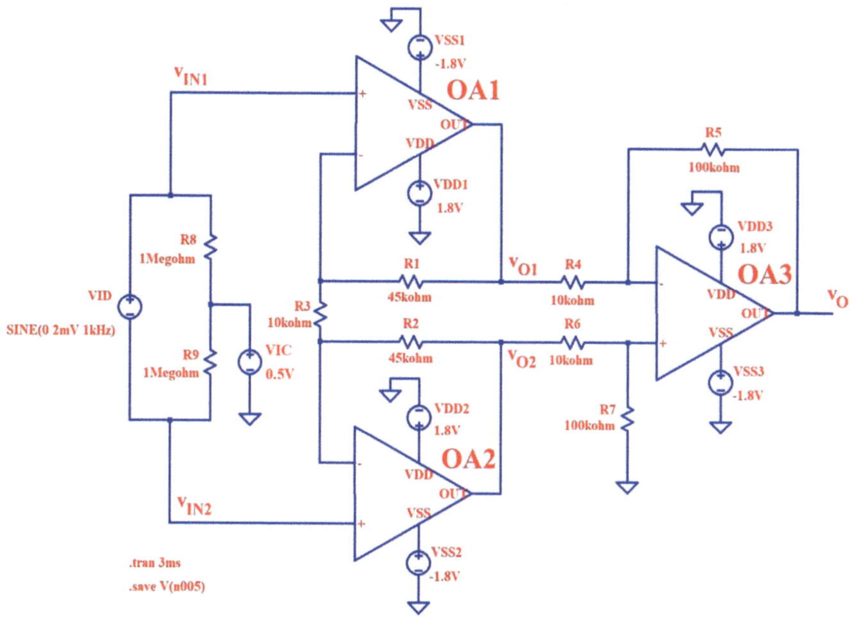

Fig. 4.51 Instrumentation amplifier circuit with a *transient analysis* in LTspice

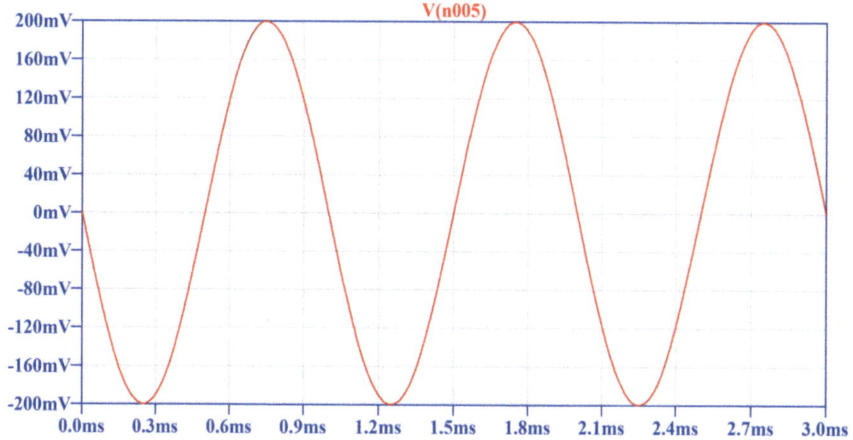

Fig. 4.52 Simulated characteristic $v_O(t)$ for the instrumentation amplifier

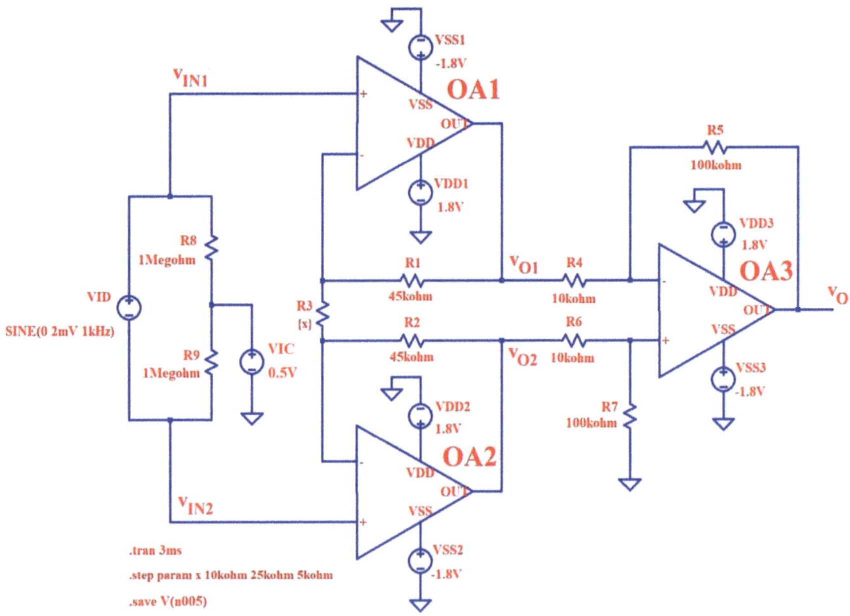

Fig. 4.53 Instrumentation amplifier with a *transient analysis* in LTspice, having as parameter the resistance R_3

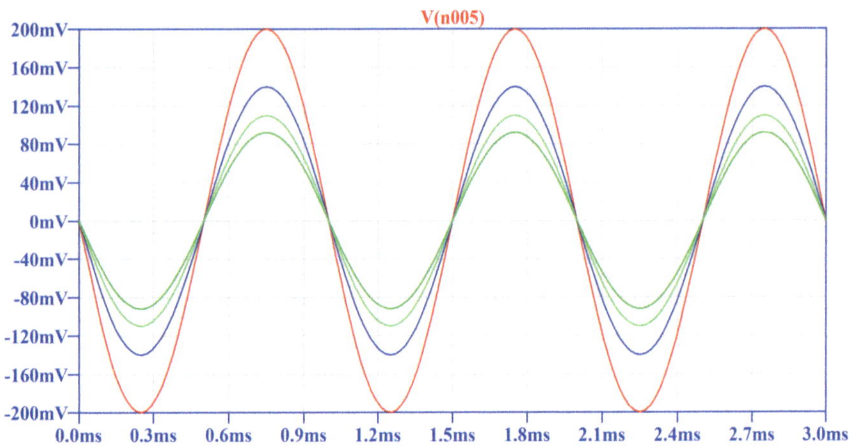

Fig. 4.54 Simulated characteristic $v_O(t)$ for the instrumentation amplifier, having as parameter the resistance R_3

4.10 Half-Wave Rectifier

A possible realization of a half-wave rectifier using an operational amplifier is presented in Fig. 4.55.

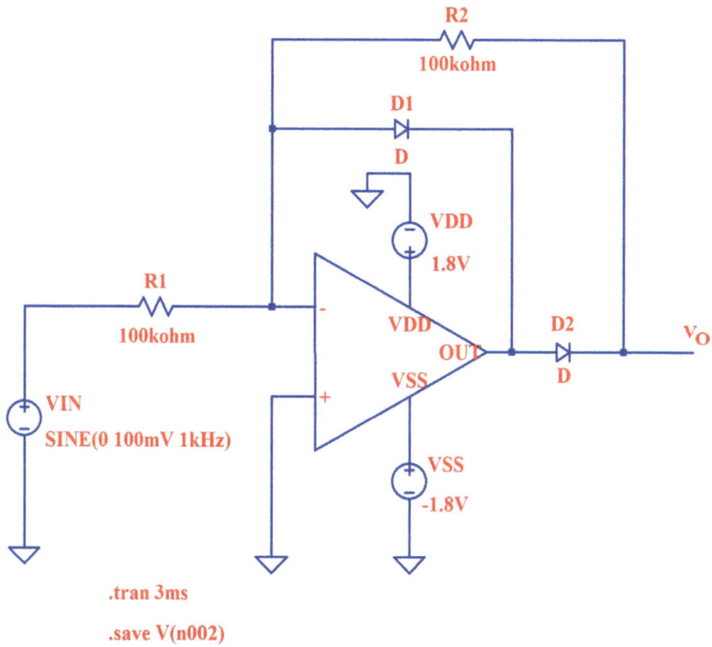

Fig. 4.55 Half-wave rectifier using an operational amplifier (*transient analysis*)

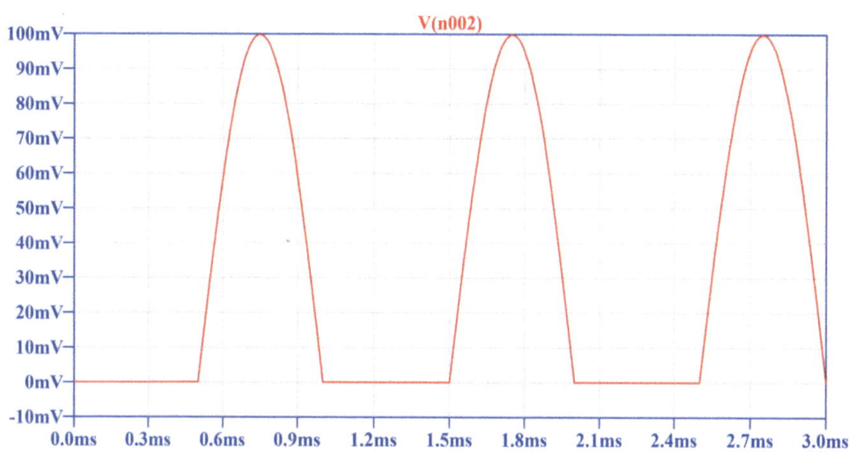

Fig. 4.56 Simulated characteristic $v_O(t)$ for the half-wave rectifier

For positive values of the input voltage v_{IN}, diode D1 is ON and diode D2 is OFF, resulting in $v_O = 0$. For negative values of the input voltage v_{IN}, diode D1 is OFF and diode D2 is ON, resulting in $v_O = -\frac{R}{R} v_{IN} = -v_{IN}$.

For graphically representing the dependence $v_O(t)$ for the half-wave rectifier from Fig. 4.55, a *transient analysis* is performed in LTspice, for a sinusoidal variation of the input voltage $v_{IN}(t)$, with an amplitude of 100 mV. In order to directly view the result of the simulation (the characteristic $v_O(t)$, shown in Fig. 4.56), the directive ".

save V(n002)" is introduced in the file, with V(n002) representing the output voltage of the circuit v_O.

4.11 Full-Wave Rectifier (Version I)

A possible realization of a full-wave rectifier (version I) using two operational amplifiers is presented in Fig. 4.57.

For positive values of the input voltage v_{IN}, diode D1 is OFF and diode D2 is ON, resulting in $v_{O1} = -\frac{R_2}{R_1} v_{IN} = -v_{IN}$. For negative values of the input voltage v_{IN}, diode D1 is ON and diode D2 is OFF. The second operational amplifier OA2 is connected as inverting voltage adder, having as inputs the voltages v_{IN} and v_{O1}. After the circuit analysis, it results that the circuit shown in Fig. 4.57 realizes the absolute value function, $v_O = |v_{IN}|$, equivalent to the implementation of a full-wave rectifier circuit.

For graphically representing the dependence $v_O(v_{IN})$ for the full-wave rectifier (version I) from Fig. 4.57, a *DC sweep analysis* is performed in LTspice, for a variation range of the input voltage v_{IN} between −100 and 100 mV. In order to directly view the result of the simulation (the characteristic $v_O(v_{IN})$, shown in Fig. 4.58), the directive ".save V(n004)" is introduced in the file, with V(n004) representing the output voltage of the circuit v_O.

For graphically representing the dependence $v_O(t)$ for the full-wave rectifier from Fig. 4.59, a *transient analysis* is performed in LTspice, for a sinusoidal variation of the input voltage $v_{IN}(t)$, with an amplitude of 100 mV. In order to directly view the result of the simulation (the characteristic $v_O(t)$, shown in Fig. 4.60), the directive ".

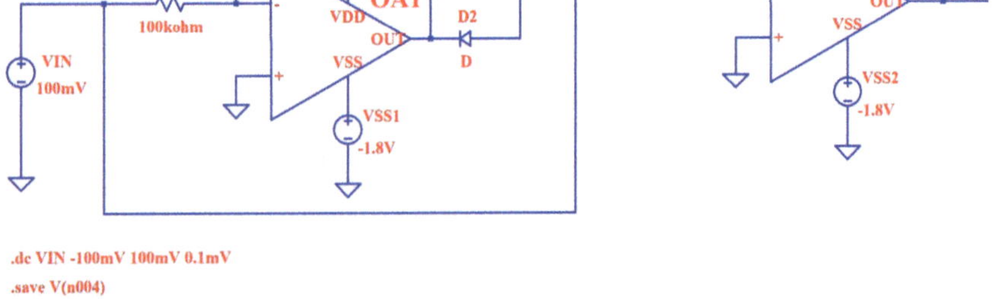

Fig. 4.57 Full-wave rectifier (version I) using two operational amplifiers (*DC sweep analysis*)

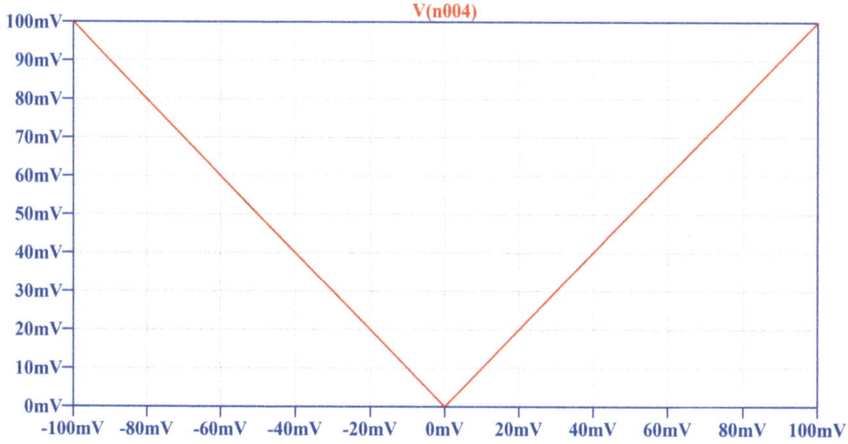

Fig. 4.58 Simulated characteristic $v_O(v_{IN})$ for the full-wave rectifier (version I)

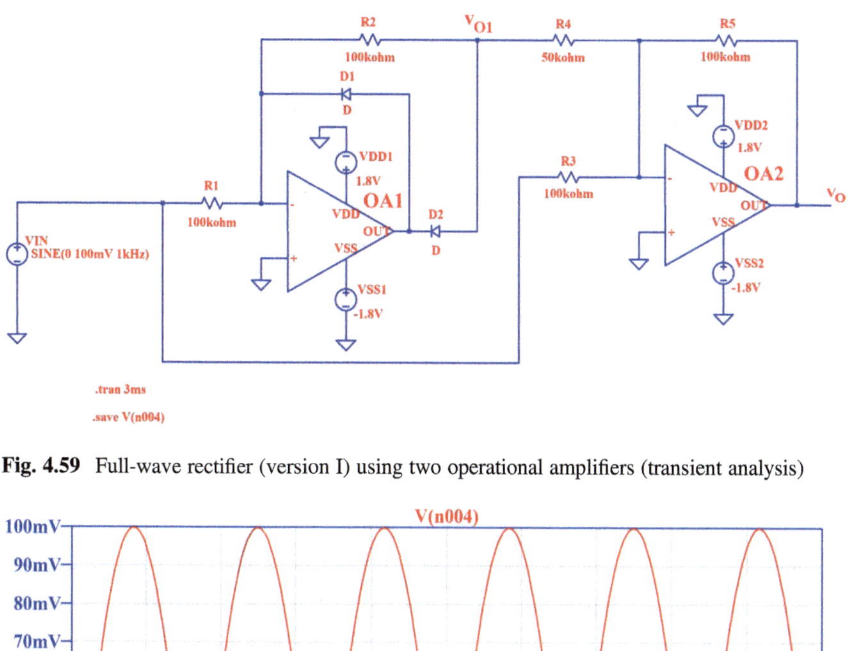

Fig. 4.59 Full-wave rectifier (version I) using two operational amplifiers (transient analysis)

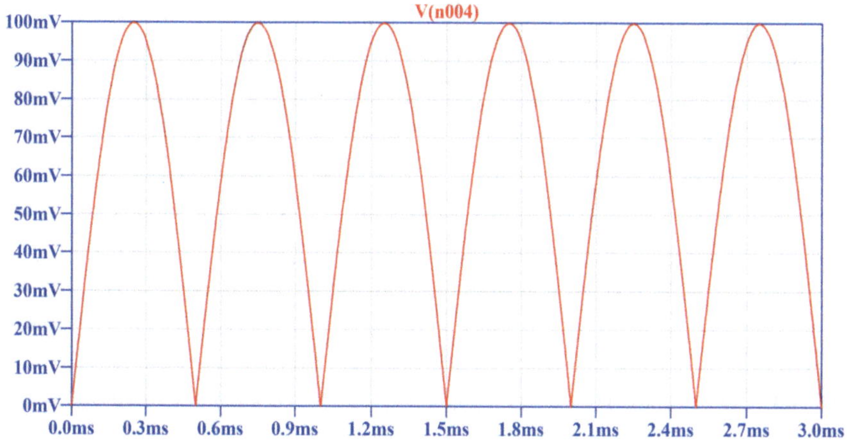

Fig. 4.60 Simulated characteristic $v_O(t)$ for the full-wave rectifier (version I)

save V(n004)" is introduced in the file, with V(n004) representing the output voltage of the circuit v_O.

4.12 Full-Wave Rectifier (Version II)

Another possible realization of a full-wave rectifier (version II) using two operational amplifiers is presented in Fig. 4.61.

For positive values of the input voltage v_{IN}, diode D1 is ON and diode D2 is OFF. For negative values of the input voltage v_{IN}, diode D1 is OFF and diode D2 is ON, resulting in $v_{O1} = \left(1 + \frac{R_2}{R_1}\right) v_{IN} = 2v_{IN}$. The second operational amplifier OA2 is connected as a difference amplifier, having as inputs the voltages v_{IN} and v_{O1}. After the circuit analysis, it results that the circuit shown in Fig. 4.61 realizes the absolute value function, $v_O = |v_{IN}|$, equivalent to the implementation of a full-wave rectifier circuit.

For graphically representing the dependence $v_O(v_{IN})$ for the full-wave rectifier (version II) from Fig. 4.61, a *DC sweep analysis* is performed in LTspice, for a variation range of the input voltage v_{IN} between -100 and $100\,\text{mV}$. In order to directly view the result of the simulation (the characteristic $v_O(v_{IN})$, shown in

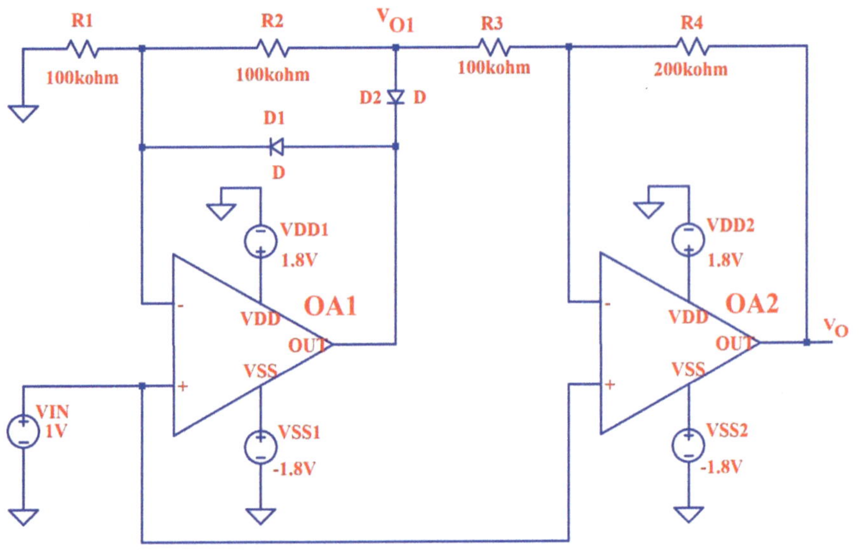

Fig. 4.61 Full-wave rectifier (version I) using two operational amplifiers (*DC sweep analysis*)

Fig. 4.62), the directive ".save V(n004)" is introduced in the file, with V(n004) representing the output voltage of the circuit v_O.

For graphically representing the dependence $v_O(t)$ for the full-wave rectifier from Fig. 4.63, a *transient analysis* is performed in LTspice, for a sinusoidal variation of the input voltage $v_{IN}(t)$, with an amplitude of 100 mV. In order to directly view the result of the simulation (the characteristic $v_O(t)$, shown in Fig. 4.64), the directive ".

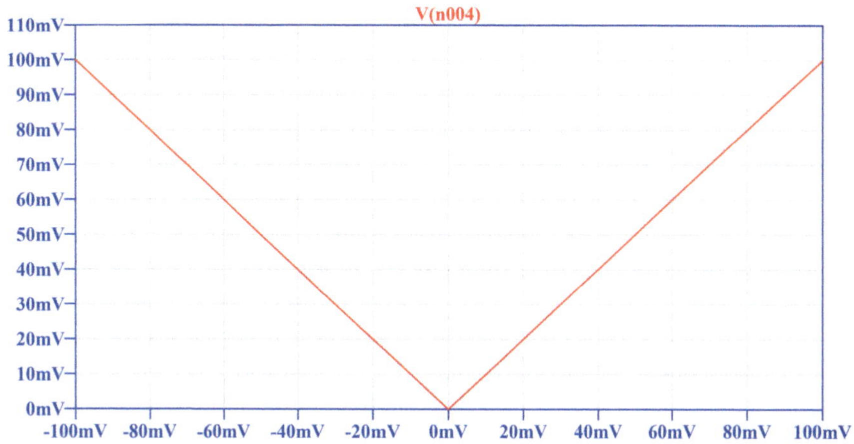

Fig. 4.62 Simulated characteristic $v_O(v_{IN})$ for the full-wave rectifier (version II)

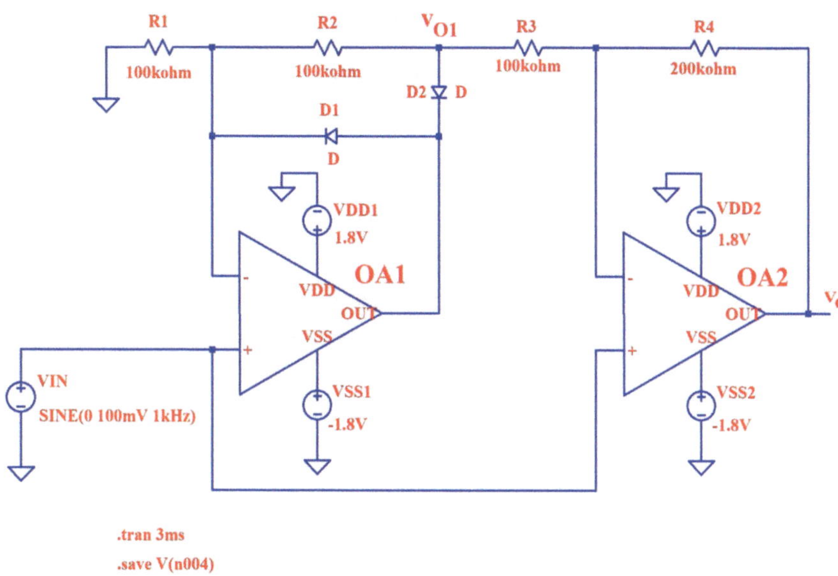

Fig. 4.63 Full-wave rectifier (version II) using two operational amplifiers (transient analysis)

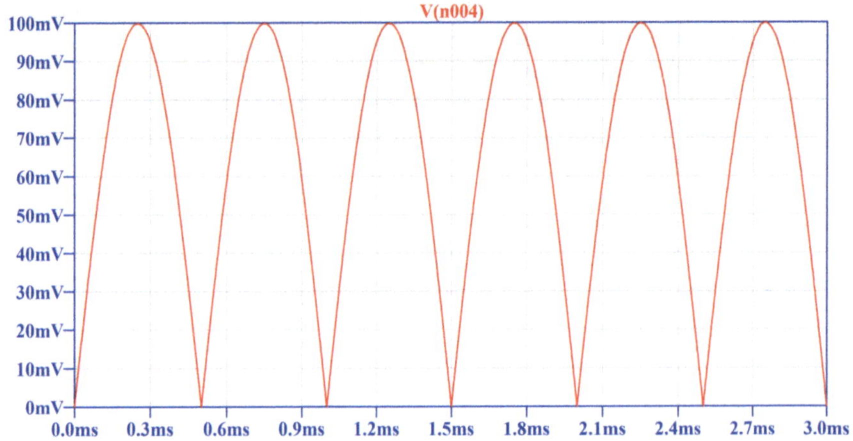

Fig. 4.64 Simulated characteristic $v_O(t)$ for the full-wave rectifier (version II)

save V(n004)" is introduced in the file, with V(n004) representing the output voltage of the circuit v_O.

References

1. Razavi B (2001) Design of Analog CMOS Integrated Circuits. New York: McGraw-Hill Education
2. Gray PR, Hurst PJ, Lewis SH and Meyer RG (2009) Analysis and Design of Analog Integrated Circuits, 5th ed. New York: Wiley
3. Allen PE and Holberg DR (2011) CMOS Analog Circuit Design, 3rd ed. New York: Oxford University Press
4. Sedra AS and Smith KC (2015) Microelectronic Circuits, 7th ed. New York: Oxford University Press

Annex

Large-Signal Model of the NMOS Transistor

The NMOS transistor is described using the *Shichman–Hodges model with channel-length modulation*. Using this model for NMOS transistor, the biasing regions and their associated mathematical expressions of the drain current I_D are considered to be the following:

Weak Inversion Region (Subthreshold Region)

$$I_D = 0, \quad \text{for} \quad V_{GS} \le V_T \tag{A.1}$$

Operation in week inversion region is not considered in this work.

Linear Region (Triode Region)

$$I_D = \frac{K'_n}{2}\left(\frac{W}{L}\right)\left[(V_{GS} - V_T)V_{DS} - \frac{V_{DS}^2}{2}\right] \times (1 + \lambda V_{DS}) \quad (\text{for } V_{GS} > V_T \text{ and } 0 \le V_{DS} \le V_{GS} - V_T). \tag{A.2}$$

C. R. Popa, *Fundamentals of CMOS Analog Integrated Circuits*,
https://doi.org/10.1007/978-3-032-30677-7

Saturation Region

$$I_D = K'_n\left(\frac{W}{L}\right)(V_{GS} - V_T)^2(1 + \lambda V_{DS}) \quad (\text{for } V_{GS} > V_T \text{ and } V_{DS} > V_{GS} - V_T), \tag{A.3}$$

where

$K'_n = \mu_n C_{ox}$ is the process transconductance parameter.
μ_n is the electron mobility.
C_{ox} is the gate oxide capacitance per unit area.
W and L are the channel width and length, respectively.
V_{GS} is the gate-source voltage.
V_{DS} is the drain-source voltage;
V_T is the threshold voltage.
λ is the channel-length modulation parameter.

The threshold voltage V_T depends on the source-bulk voltage V_{SB}, with this mathematical dependence being modelled by the following relation:

$$V_T = V_{T0} + \gamma\left(\sqrt{V_{SB} + |2\Phi_F|} - \sqrt{|2\Phi_F|}\right), \tag{A.4}$$

where V_{T0} is the threshold voltage without biasing the bulk terminal ($V_{SB} = 0$), γ represents the bulk threshold parameter (or body effect constant), while $|\Phi_F|$ is the Fermi potential of the body. This second-order effect is named body effect.

The model adopted in this work reflects a deliberate trade-off between physical accuracy and analytical tractability:

- Preserves the square-law dependence, enabling closed-form solutions.
- Includes channel-length modulation effect, essential for:
  - Finite output resistance.
  - Gain estimation.
  - Current mirror accuracy.
- Avoids higher-order effects that would complicate hand analysis.

This level of abstraction is sufficient to capture the dominant behavior of the circuits studied in this book.

The following effects are neglected:

- Body effect.
- Mobility degradation.
- Velocity saturation.
- Short-channel effects.

While the *Shichman–Hodges model* with channel-length modulation effect provides sufficient accuracy for hand analysis and conceptual understanding, modern MOS transistors (e.g. BSIM) incorporate numerous short-channel and second-order effects to achieve high accuracy in deep-submicron technologies. Such models, however, rely on complex parameter sets and numerical evaluation, making them less suitable for transparent analytical treatment.

A simple model, applied rigorously, is preferred over a complex model that obscures insight. This enables transparent derivations, intuitive understanding and direct comparison between theory and simulation.

Small-Signal Model of the NMOS Transistor

The following simplest model of the NMOS transistor is used in this work (Fig. A.1):

For circuit analysis, the following small-signal quantities are defined:

Transconductance:

$$g_m = \frac{\partial I_D}{\partial V_{GS}} = K'_n\left(\frac{W}{L}\right)(V_{GS} - V_T)(1 + V_{DS}) \cong \sqrt{2K'_n\left(\frac{W}{L}\right)I_D}. \tag{A.5}$$

Drain-source resistance:

$$rds = \frac{1}{g_{ds}} = \frac{1}{\frac{\partial I_D}{\partial V_{DS}}} = \frac{1}{\lambda\frac{K'_n}{2}\left(\frac{W}{L}\right)(V_{GS} - V_T)^2} \cong \frac{1}{\lambda I_D}. \tag{A.6}$$

Because the gate current is considered to be zero for this model, the equivalent resistance in the gate of a transistor will be infinite.

Fig. A.1 Model of NMOS transistor

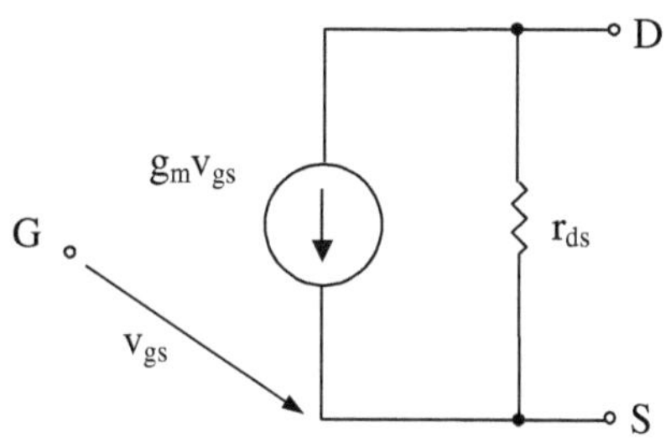

Dynamic Resistances

In order to determine the general expression of the small-signal resistance seen looking into the drain of an NMOS transistor (very useful for avoiding complicate computations when calculating the output resistance of a current source or of an amplifier), the following general circuit can be considered (Fig. A.2):

The small-signal resistance seen looking into the drain of the transistor can be determined as follows:

$$v_x = \left(i_x - g_m v_{gs}\right) r_{ds} + i_x R_2, \tag{A.7}$$

$$v_{gs} = -i_x R_2, \tag{A.8}$$

resulting in

$$R_D \stackrel{def}{=} \frac{v_x}{i_x} = r_{ds}(1 + g_m R_2) + R_2. \tag{A.9}$$

So, the small-signal resistance seen looking into the drain of the transistor will have the following expression:

$$R_D = r_{ds}(1 + g_m R_2). \tag{A.10}$$

The equivalent resistance R_D is strongly increased by the negative feedback introduced by resistance R_2.

In order to determine the general expression of the small-signal resistance seen looking into the source of an NMOS transistor, the following general circuit can be considered (Fig. A.3):

The small-signal resistance seen looking into the source of the transistor can be determined as follows:

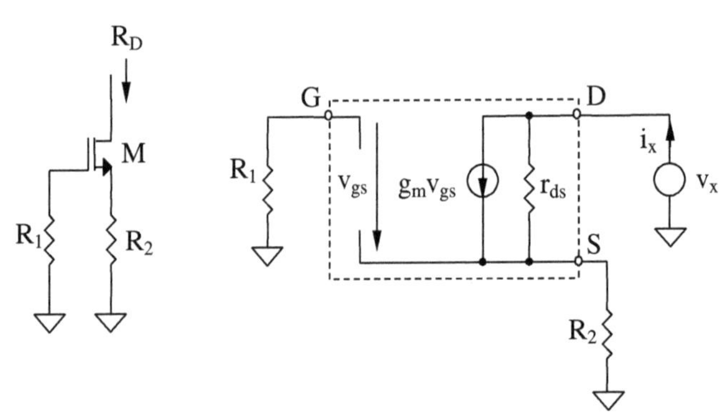

Fig. A.2 General circuit for determining the small-signal resistance seen looking into the drain

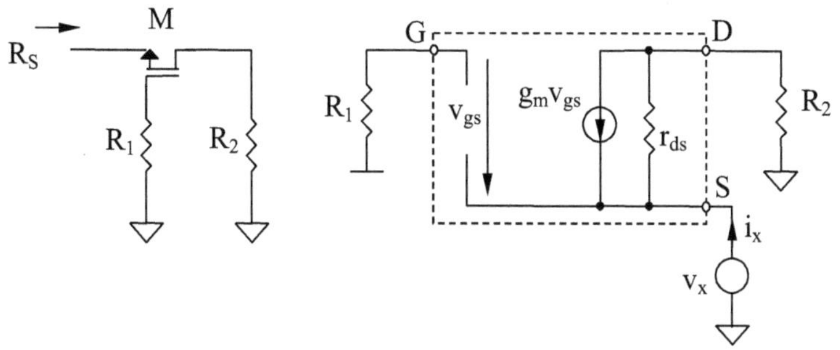

Fig. A.3 General circuit for determining the small-signal resistance seen looking into the source

$$i_x \cong -g_m v_{gs} = g_m v_x, \tag{A.11}$$

resulting in the following expression R_S:

$$R_S \stackrel{def}{=} \frac{v_x}{i_x} = \frac{1}{g_m}. \tag{A.12}$$

Simulation Framework

All analytical results are verified using LTspice. Simplified MOS models are employed:

.model cmosn NMOS (Kp = 200 μ Vto = 0.5 V lambda = 0.01)

.model cmosp PMOS (Kp = 100 μ Vto = −0.5 V lambda = 0.02).

Supply voltages: $V_{DD} = 1.8\,\text{V}$ and $V_{SS} = -1.8\,\text{V}$.

Three fundamental types of analyses available in LTspice are used in this work, exemplified here using the basic inverting amplifier.

The .dc Analysis (DC Sweep Analysis)

It is used to determine, for example, the static transfer characteristic of the basic inverting amplifier (Fig. A.4). In this case, the input voltage source v_{IN} from Fig. A.4 is swept over a specified range, between −2 mV and 2 mV, and the corresponding output voltage is computed. This allows the visualization of the input-output relationship $v_O(v_{IN})$, highlighting the linear operating region and allowing to determine the voltage gain $\frac{v_O}{v_{IN}} = -\frac{R_2}{R_1} = -10$, equal to the slope of the graphical characteristic $v_O(v_{IN})$.

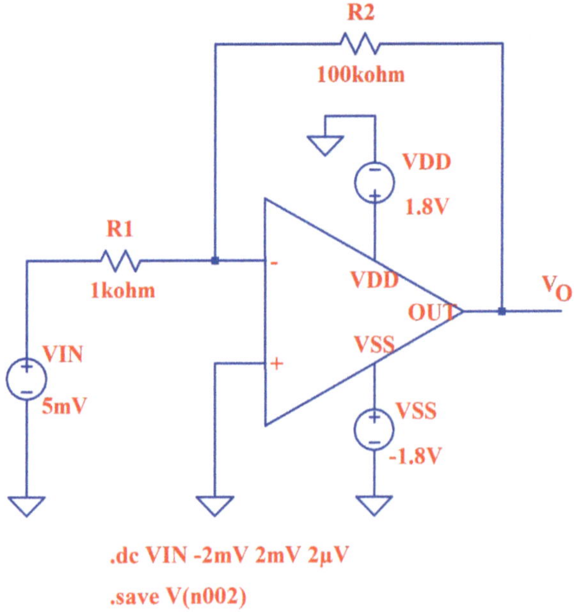

Fig. A.4 Basic inverting amplifier with a **.dc analysis (DC sweep analysis)** in LTspice

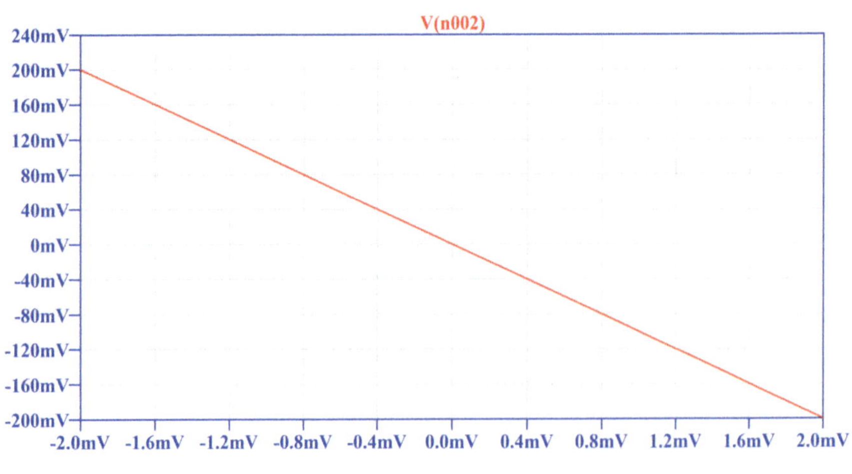

Fig. A.5 Simulated characteristic $v_O(v_{IN})$ for the basic inverting amplifier

The simulated characteristic $v_O(v_{IN})$ for the basic inverting amplifier is shown in Fig. A.5. In order to directly view the result of the simulation (the characteristic $v_O(v_{IN})$, shown in Fig. A.5), the directive ".save V(n002)" is introduced in the file, with V(n002) representing the output voltage of the circuit v_O.

The.step param Analysis

It is used for illustrating the influence of a parameter on the operation of a circuit. For example, for the basic inverting amplifier, in order to illustrate the influence of the resistance R_2 on the voltage gain of the circuit, the previous **.dc analysis (DC sweep analysis)**—for a variation of the voltage v_{IN} between $-2\,\text{mV}$ and $2\,\text{mV}$—is complemented by a **.step param analysis** (for $x = R_2$, between $40\,\text{k}\Omega$ and $100\,\text{k}\Omega$, with a step of $20\,\text{k}\Omega$). The circuit used for this simulation is shown in Fig. A.6.

The simulated characteristic $v_O(v_{IN})$ for the basic inverting amplifier, having as parameter the resistance R_2, is shown in Fig. A.7. In order to directly view the results of the simulation (the characteristics $v_O(v_{IN})$, shown in Fig. A.7), the directive ".save V(n002)" is introduced in the file, with V(n002) representing the output voltage of the circuit v_O.

The.tran Analysis (Transient Analysis)

It is used to study the time-domain response of the circuit to a time-varying input signal. For the inverting amplifier, a sinusoidal input is applied, and the resulting output waveform is observed. This analysis provides insight into dynamic behavior,

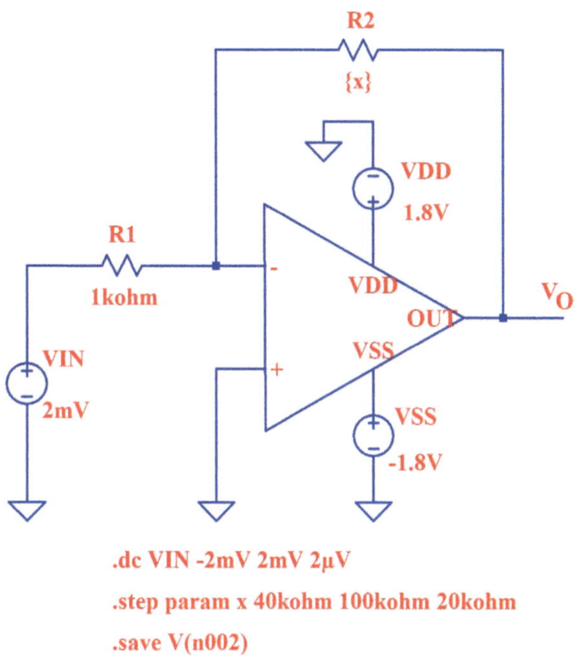

Fig. A.6 Basic inverting amplifier with a **.dc analysis (DC sweep analysis)** in LTspice, having as parameter the resistance R_2 (a **.step param analysis**)

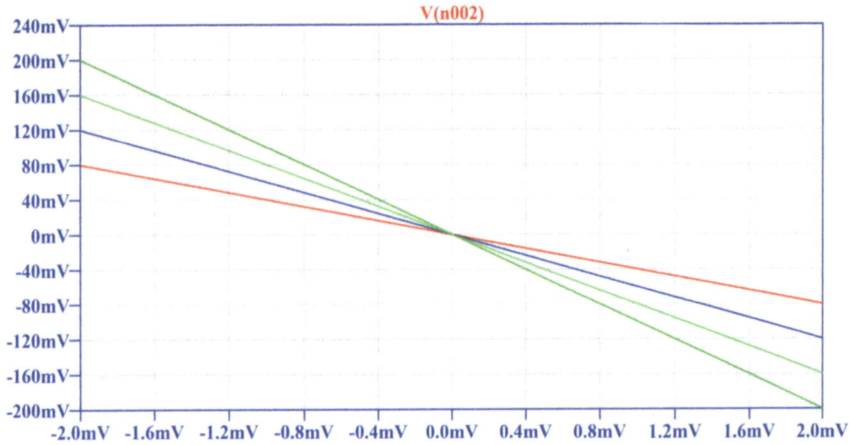

Fig. A.7 Simulated characteristic $v_O(v_{IN})$ for the basic inverting amplifier, having as parameter the resistance R_2

Fig. A.8 Basic inverting amplifier circuit with a **.tran analysis (transient analysis)** in LTspice

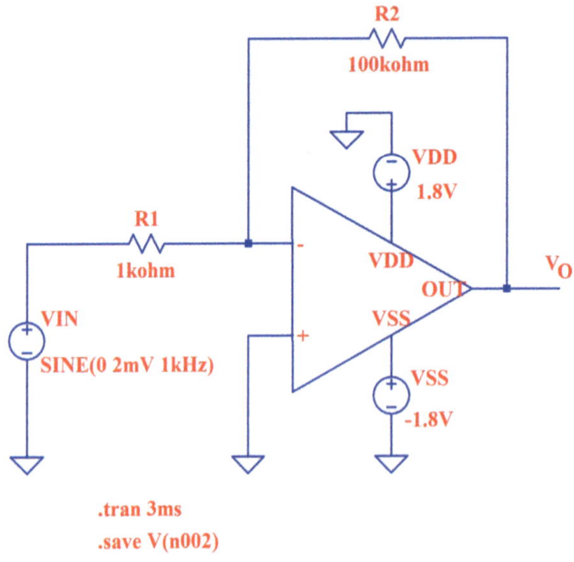

including signal amplification, phase inversion and possible distortion caused by the large-signal effects or limited output swing. The circuit used for this simulation is shown in Fig. A.8.

The simulated characteristic $v_O(t)$ for the basic inverting amplifier is shown in Fig. A.9. In order to directly view the result of the simulation (the characteristic $v_O(t)$, shown in Fig. A.9), the directive ".save V(n002)" is introduced in the file, with V (n002) representing the output voltage of the circuit v_O.

Fig. A.9 Simulated characteristic $v_O(t)$ for the basic inverting amplifier

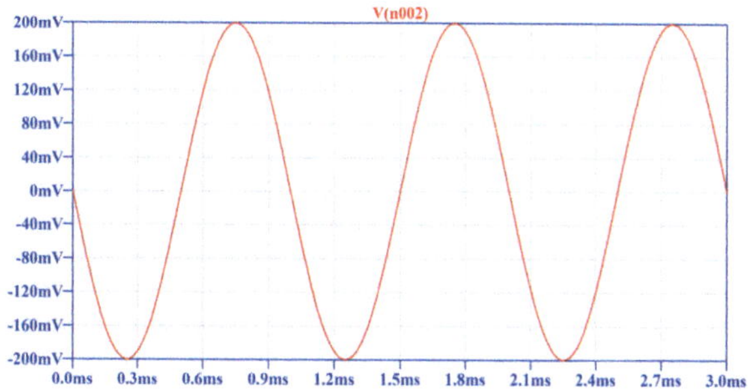

MIX
Papier aus verantwortungsvollen Quellen
Paper from responsible sources
FSC® C105338

If you have any concerns about our products,
you can contact us on
ProductSafety@springernature.com

In case Publisher is established outside the EU,
the EU authorized representative is:
Springer Nature Customer Service Center GmbH
Europaplatz 3, 69115 Heidelberg, Germany

Printed by Libri Plureos GmbH
in Hamburg, Germany